The Whens and Wheres of a Scientific Life

Global Science Education

Series Editor
Professor Ali Eftekhari

Learning about the scientific education systems in the global context is of utmost importance now for two reasons. Firstly, the academic community is now international. It is no longer limited to top universities, as the mobility of staff and students is very common even in remote places. Secondly, education systems need to continually evolve in order to cope with the market demand. Contrary to the past, when the pioneering countries were the most innovative ones, now emerging economies are more eager to push the boundaries of innovative education. Here, an overall picture of the whole field is provided. Moreover, the entire collection is indeed an encyclopaedia of science education and can be used as a resource for global education.

The Whens and Wheres of a Scientific Life

John R. Helliwell

CRC Press
Taylor & Francis Group
Boca Raton London New York

CRC Press is an imprint of the
Taylor & Francis Group, an **informa** business

First edition published 2021
by CRC Press
6000 Broken Sound Parkway NW, Suite 300, Boca Raton, FL 33487-2742

and by CRC Press
2 Park Square, Milton Park, Abingdon, Oxon, OX14 4RN

Library of Congress Cataloging-in-Publication Data
Names: Helliwell, John R., author.
Title: The whens and wheres of a scientific life / John R. Helliwell.
Description: First edition. | Boca Raton : CRC Press, 2021. | Series:
Global science education | Includes bibliographical references and index.
Identifiers: LCCN 2020047276 (print) | LCCN 2020047277 (ebook) |
ISBN 9780367489717 (hardback) | ISBN 9781003043744 (ebook)
Subjects: LCSH: Helliwell, John R. | Science--Philosophy. |
Scientists—Great Britain--Biography. | Chimists—Great
Britain—Biography. | Crystallographers—Great Britain--Biography.
Classification: LCC Q175 .H394585 2021 (print) | LCC Q175 (ebook) |
DDC 501—dc23
LC record available at https://lccn.loc.gov/2020047276
LC ebook record available at https://lccn.loc.gov/2020047277

ISBN: 9780367489717 (hbk)
ISBN: 9781003043744 (ebk)

Typeset in Minion
by codeMantra

Contents

Adapted from Samuel Taylor Coleridge (1772–1834):

In Xanadu did Kubla Khan
A stately pleasure-dome decree:
Where Alph, the sacred river, ran
Through caverns measureless to man
Down to a sunlit sea.
So twice five miles of fertile ground
With walls and towers were girdled round;
And there were gardens bright with sinuous rills,
Where blossomed many an incense-bearing tree;
And here were forests ancient as the hills,
Enfolding sunny spots of greenery.

......

......

And close your eyes
For we on honey-dew hath fed,
And drunk the milk of Paradise.

Preface

ON FINISHING MY THREE PREVIOUS BOOKS in which I explored the *hows*, the *whys* and the *whats* of the scientific life, I thought, 'great, a nice box set'! But then, my book publisher Hilary Lafoe remarked to me, the '*whens* and the *wheres*' would be a natural complement to the previous three. There are of course so many *whens* and *wheres*. *When* and *where* questions play a big role in major science facility decisions. *When* and *where* play a big role in controlling a pandemic like the coronavirus COVID-19. In one of my chapters, I describe the quantum world, which does not actually allow an electron's position to be known precisely *where* and *when* together; I describe the profound consequences of this for daily life.

In another aspect of one's scientific life, the decisions one takes can be due to *when* or *where* you find yourself or both. In the very recent TV movie *Devs*[1] about quantum computing being so powerful compared with digital computing, it was argued that since any effect is due to a cause, it should be possible to recreate any scene from the past or predict the future. In 1999, I considered the future of biological crystal structure analysis as a lecturer in a symposium on 'Next Century Biophysics'. A fellow speaker, Professor Charles Cantor, described a 'what if' scenario in which computing could become sufficiently powerful to predict experiments without doing them. When considering the *Devs* vision

---

[1] https://en.wikipedia.org/wiki/Devs.

and the Cantor vision in the context of one's career, there is sometimes a cause and effect feel to it, but our imaginations and ideas can fundamentally change that.

Also big questions and issues arise about the role of the scientific life in our society and in our world. These have to do with trusting science at all or with the wider roles of the scientist. Such questions arose in my mind while reviewing new books that I took on or in considered reflection as my scientific life lengthened, i.e. into my semi-retirement. I have also included several biographical sketches that I have written of some of those people who directly influenced me. Three were British. These included Dame Kathleen Lonsdale a physicist and turned chemist, whose life involved her efforts for global peace. For an international perspective, I have included a biographical sketch of a leading scientist in India, Professor MaMannaMana Vijayan, a physicist who steadily became a biologist, who eventually became the President of the Indian National Academy of Sciences.

Acknowledgements

I AM GRATEFUL TO ALL who I have worked with, or who have supported, my scientific life as well as to my work places, notably the University of Manchester where I have worked as a professor of structural chemistry these last 30 years, often jointly with the Scientific and Technical Facilities Council facility, Daresbury Laboratory. Likewise, I have had various interactions with different organisations in the chapters of this book, especially the International Union of Crystallography. I also thank the British Crystallographic Association, the European Crystallographic Association, the American Crystallographic Association and the British Biophysical Society for the community fora that they provided. I am grateful to all who shaped my training and my career.

Some organisations are not or are only briefly mentioned but have also had a great influence in my scientific life. I have not for example detailed my work on the synchrotron or neutron facility committees around the world on which I have served as a member or chair. I have not described my various roles in the British Crystallographic Association or my work when I was President of the European Crystallographic Association. I have also not described my work for the Institute of Physics Panels I served on assessing many Fellowship applications, nor my work in recent years for the American Institute of Physics Publishing and the American Crystallographic Association.

The drafts of this book were very helpfully improved by Emeritus Professor Gervais Chapuis of the Ecole Polytechnique Fédérale in Lausanne, Katrine Bazeley MRVS and Brian McMahon of the IUCr.

I am also indebted to all my PhD students and postdocs over a period of 30 years, listed on this plaque (Figure A.1) presented to me in Albuquerque, New Mexico, in 2014 at the American Crystallographic Association Annual Conference on the occasion of my receiving the A. L. Patterson Award. It was compiled by my former PhD student, Dr Dora Gomez, of the University of the Andes, Merida, Venezuela, now retired to Panama.

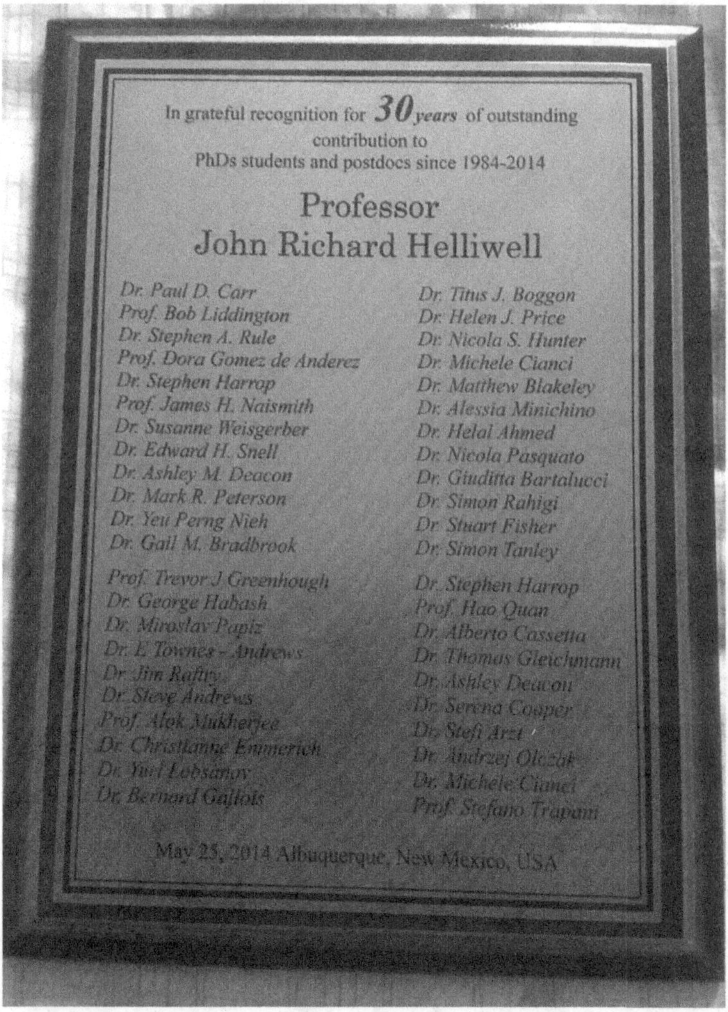

FIGURE A.1   Plaque presented to me in Albuquerque, New Mexico, in 2014 at the American Crystallographic Association (ACA) Annual Conference on the occasion of my receiving the ACA A. L. Patterson Award. It was compiled by my former PhD student, Dr Dora Gomez.

Author

John R. Helliwell, DSc (Physics, University of York), DPhil (Molecular Biophysics, Oxford University), is Emeritus Professor of Chemistry at The University of Manchester, where he served as Professor of Structural Chemistry from 1989 to 2012. Academic teaching from 1979 till 1988 was at the Universities of Keele and York in the physics departments there. He is a researcher in the fields of crystallography, biophysics, structural biology, structural chemistry and data science. He was also based at the Synchrotron Radiation Source at the UK's Daresbury Laboratory, in various periods of appointment between 1979 and 2008, including in 2002 as Director of Synchrotron Radiation Science. He is a Fellow of the Institute of Physics, the Royal Society of Chemistry, the Royal Society of Biology, and the American Crystallographic Association, and an Honorary Member of the British Crystallographic Association and of the British Biophysical Society. He is a Corresponding Member of the Royal Academy of Sciences and Arts of Barcelona, Spain and Honorary Member of the National Institute of Chemistry, Slovenia. His awards include the European Crystallographic Association Eighth Max Perutz Prize 2015, the American Crystallographic Association Patterson Award 2014 and the 'Professor K Banerjee Endowment Lecture Silver Medal' of the Indian Association for the Cultivation of Science (IACS) 2001. He published over 200 scientific research papers and several books, e.g. *Macromolecular Crystallography with Synchrotron Radiation* with

Cambridge University Press (1992), published in paperback in 2005 and *Macromolecular Crystallization and Crystal Perfection* with N. E. Chayen and E. H. Snell, Oxford University Press – International Union of Crystallography Monographs on Crystallography (2010). He has published several *Scientific Life*, popular science, books in recent years, which are with CRC Press/ Taylor & Francis Group.

I

The Development of My Scientific Life

CHAPTER 1

When and Where I Embraced the Scientific Life[1]

I WAS THE FIRST IN my family to be interested in science and the second to go to university. At school in West Yorkshire, my best subjects were history, geography and mathematics, but at age 15 in the UK, I had to choose between humanities or science. I chose to specialise in chemistry, mathematics and physics, having given up biology, as I was too squeamish even to dissect a worm. My father was a policeman, and he was moved around rather a lot; I had four schools between the age of 11 and 18 years. These were Todmorden Grammar School, Thornes House School Wakefield, Price Henry's Grammar School Otley and Ossett Grammar School, all in West Yorkshire. I had to develop a

[1] This is a modified extract of my Biographical Memoir, prompted by my A. L. Patterson Prize Award from the American Crystallographic Association in 2014 for the development of crystallographic methods. This section was first published in the *RefleXions* magazine of the American Crystallographic Association, and my article is also available online at the ACA's History Portal (https://history.amercrystalassn. org/).

self-reliance both catching up on school course notes and always making new friends. My mother was a nurse. I was an only child. I played for my school at rugby, and we travelled most Saturdays in the winter to different schools in West Yorkshire. I went to York University to read for a physics degree. It was a small class size (32 students) and this was important, as I was always able to do the physics undergraduate laboratory practicals on my own, which I especially felt was important to my training as a physicist.

My DPhil (1974–1977) supervisor in the Laboratory of Molecular Biophysics at Oxford University was Dr Margaret Adams, who had been a student of Professor Dorothy Hodgkin (Nobel Prize in Chemistry, 1964); I was Margaret's first DPhil student. Professor Charlie Bugg of the University of Alabama, later to become a biotechnologist as Chief Scientific Officer of Biocryst Pharmaceuticals, and a pioneer of protein crystallisation in space, was a Visiting Scientist with Margaret through my first year. He was also my proposer for my membership of the American Crystallographic Association (ACA), and I was duly admitted on 13 November 1975. My interest in synchrotron radiation (SR) arose early on during my DPhil as I thought that the various experimental challenges for macromolecular crystallography could be handled better. In my DPhil project, our crystals were quite typical and showed weak diffraction, requiring long exposure times on our rotating anode X-ray source. Solving the crystallographic phase problem, the crucial step in determining a molecular structure, seemed to me haphazard with the techniques then available to us. As a graduate member of the UK Institute of Physics, I heard that the Nobel Prize winning physicist Mössbauer was to give a lecture at the Rutherford Appleton Laboratory, near Oxford, and during his talk, he suggested the use of nuclear anomalous dispersion to solve the crystallographic phase problem but was technically very difficult. Meanwhile, again while I was a DPhil student, the IUCr conference book *Anomalous scattering* was published, which included a very stimulating chapter by Hoppe and Jakubowski [1]. With other colleagues around the world, this pioneering experiment with home laboratory X-ray tubes was greatly improved on and brought to general use using the upcoming synchrotron light facilities, although it took a decade or so of developments. Suffice to say, these examples confirmed my view that it was a fertile time for a physicist like me entering protein crystallography – although at my interview, one person said, 'don't bother, all the methods are fine as they are, there is no place for a physicist!'

Dorothy Hodgkin won the Nobel Prize for Chemistry because of her groundbreaking work in determining the structure of many biologically important molecules. She informed me that she had received news (from Professor Sir Ron Mason) about some developments involving the first protein crystallography experiments at the Stanford SSRL, led by professor of chemistry Keith Hodgson. I had the very good fortune to work closely with Keith some 10 years later along with Dr Britt Hedman, staff scientist of the SSRL. Dorothy asked my opinion of the SSRL work and I reported back to her that I found the preprint that she had passed on to me very exciting! The opening sentence of the article by Keith and his coworkers published shortly after [2] was 'The use of synchrotron radiation as a source for single crystal X-ray diffraction studies has recently been the subject of considerable discussion and controversy.' The first reference in this paper was to pioneering work at DESY in Hamburg on the use of SR in the biological diffraction of (predominantly) muscle fibres [3]. This I also found an exciting paper. The DESY acronym means "Deutsches Elektronen-Synchrotron" (English *German Electron Synchrotron*) and is a national research centre in *Germany* that operates *particle accelerators* used to investigate the structure of matter. My early curiosity about the use of SR stimulated my subsequent career very significantly.

So, as a DPhil student I thought a lot about 'How to solve the phase problem?' A way to get the phase of a reflection was using resonant X-ray scattering: two wavelengths and an anomalous difference were needed. In my DPhil thesis [4], which was concerned with the determination of the molecular structure of a biologically important enzyme, I included an Appendix on my efforts at the Northern Institutes Nuclear Accelerator (NINA) synchrotron (Daresbury Laboratory, Figure 1.1) in 1976 to use the anomalous scattering method that were then so novel and exciting [1,2]. I also had tried to measure diffraction data at NINA on small crystals of the protein despentapeptide insulin, with Dr Guy Dodson's help. Incidentally, Guy was quite

FIGURE 1.1 Prime Minister Harold Wilson (1916–1995; MP for Huyton, near, 14 miles from, Daresbury, 1945–1964) departing Daresbury Laboratory in June 1967 having formally opened it. This is the entrance to B Block. My office 'B12', when I was an SRS beamline scientist, was on the ground floor just down the corridor from here. When I later became Director of SR Science at the CCLRC my office was B67, and my PA Marie White office was B67A, on the first floor of this building. The first synchrotron built there was called NINA, the Northern Institutes Nuclear Accelerator (see http://www.synchrotron. org.uk/index.php?option=com_content&view=article&id=58:ian-munros-memories&catid=52:reminiscences&Itemid=55). This photo kindly provided by STFC Daresbury Laboratory).

merciless in making fun of my Yorkshire accent, which was very strong at that time. My local contact at NINA was Dr Joan Bordas, who much later (in ~2010), when he was Director of the Spanish synchrotron radiation source ALBA, invited me to chair their Science Advisory Committee and be the president of their Beamtime Panels. As in any scientific life, the collaborations and connections made at any stage of one's career will prove to be very important as that career develops.

It was during my DPhil that I met my wife to be, Madeleine (*née* Berry), in Holywell Manor, the joint Balliol and St Anne's Colleges Graduate Centre, where we were both resident. Madeleine was doing her DPhil with Professor Malcolm Green FRS in synthetic inorganic chemistry. We married in 1978. She had several postdoctoral chemistry research posts. Later, after a career break having our three children, she retrained and became a chemical crystallographer firstly at York University, at the initiative of Guy who had moved to York University and then for many years in Manchester University. Madeleine and I have published about ten papers together, one of which I highlighted in my Patterson Award Lecture; she has enjoyed a successful career, with about 400 publications altogether.

After I completed my DPhil, I embarked on postdoctoral research with Margaret in Oxford, funded by the Medical Research Council. I also won a Junior Research Fellowship at Linacre College. But within a few months, I was offered a joint appointment at Keele University and at the Synchrotron Radiation Source (SRS) under construction at the Daresbury Laboratory, which seemed to me to be an especially exciting research opportunity as I have emphasised in describing my doctoral research but also because it was a 5-year post allowing more career continuity. Although the protein crystallography community seemed sceptical about the future role of SR, I was able to obtain UK community support to establish the first and subsequent instruments for protein crystallography at the SRS, and then at the European Synchrotron Radiation Facility (ESRF) in

FIGURE 1.2 Aerial view of the European Synchrotron Radiation Facility and the Institut Laue Langevin Neutrons Reactor Source in Grenoble, France. This photograph was kindly provided by Delphine Chenevier, Head of ESRF Communications, Grenoble, France and reused here with her permission.

Grenoble (Figure 1.2) [5,6]. ESRF was the first of the third generation of high-brilliance SR sources based on X-ray undulators and initiated a global revolution in X-ray science. Undulators are also at the core of the subsequent X-ray laser source developments.

REFERENCES

1. Hoppe, H. and Jakubowski, V. (1975). The determination of phases of erythrocruorin using the two-wavelength method with iron as anomalous scatterer. In *Anomalous Scattering*, edited by S. Ramaseshan and S. C. Abrahams, pp. 437–461. Copenhagen: Munksgaard.
2. Phillips, J.C., Wlodawer, A., Yevitz, M.M. and Hodgson, K.O. (1976). Applications of synchrotron radiation to protein crystallography: Preliminary results. *Proc. Natl. Acad. Sci. USA* **73**, 128–132.
3. Rosenbaum, G., Holmes, K.C. and Witz, J. (1971). Synchrotron radiation as a source for X-ray diffraction. *Nature* **230**, 434–437.

4. Helliwell, J.R. (1977). X-ray studies concerning the structure of 6-phosphogluconate dehydrogenase. DPhil Thesis, University of Oxford, UK. Digital copy available from the Bodleian Library. https://ora.ox.ac.uk/objects/uuid:88b2605f-6ad4-48b2-bbd9-94b4d2661fc3.

5. Helliwell, J.R. (1992). *Macromolecular Crystallography with Synchrotron Radiation.* Cambridge University Press, Cambridge, UK. Published in paperback 2005.

6. Branden, C.-I. (1994). The new generation of synchrotron machines. *Structure* 2, 5–6.

CHAPTER 2

When Did I Become a Crystallographer?

I N PART, THE QUESTION of when I became a crystallographer is answered in my short memoir in Chapter 1. But the *precise* moment is difficult to pin down, exacerbated as well because a precise definition of crystallography is difficult to come by. It is a very broad subject that overlaps with almost any interest in the material world.

Emeritus Professor André Authier proposed (p. 129 of his book [1]) a definition of ourselves as crystallographers. Paraphrasing only slightly:

> What is it that keeps the clan of crystallographers together? The inner structure of a material, ordered or not, its imperfections, at the nanoscopic, microscopic, or macroscopic scales, are directly related with its physical, chemical, mineralogical, or biological properties. The common goal of unravelling these structures or their defects by a large variety of techniques, and understanding their relations to the properties of materials, is what keeps the crystallographers together.

Thus, though you can grow the crystallographer out of physics or chemistry or biology, can you take the physicist, chemist or biologist out of the crystallographer? This is a well-known and recurrent question. William Lawrence Bragg (Nobel Prize in Physics 1915, with his father William Henry Bragg, awarded for their services in the analysis of crystal structure by means of X-rays) in both Manchester and Cambridge faced the question: Is crystallography proper physics? I know that the Physics Department in Manchester were proud enough of their former Langworthy Professor of Physics to invite me to deliver the W. L. Bragg Lecture on the occasion of the University's 150th Anniversary[1]. We 'crystallographers as a clan', as André Authier calls us, need to be careful, though, as we are a science subject without undergraduate portfolio. Crystallography gets only crumbs of space within the crowded curricula of physics or chemistry or biochemistry degree courses. The national, regional and international crystallography bodies needed to take, as they still do, the firmest of roles in furthering crystallographic education of scientists from all disciplines.

So, why is crystallography important? i.e. that made me wish to focus on it so much as a research area. People in general have a curiosity about crystals. I have taken the time and trouble to meet the public and schoolchildren at our University's School of Chemistry Open Days; explaining to the media our research results; giving open lectures to workers' associations and the 'university of the third age'; and in schools themselves giving 'general science interest' lectures. I had the pleasure of even delivering a Royal Institution Friday Evening Discourse, Baroness Greenfield presiding, on the topic 'Why does a lobster change colour on cooking?' (see Chapter 19). Explaining crystallography is a joy I find. I like to start with showing people the double image effect in calcite and ask 'how do you think that happens?' Besides natural curiosity, at the other end of the spectrum, are the utility of

---

[1] Available at https://www.iucr.org/education/ teaching-resources/bragg-lecture-2001.

our discoveries as a field. The discovery of new pharmaceuticals for disease treatments is also very much assisted these days by structure based drug design, which has at its core crystallography; a general description of this is by Professor Charlie Bugg (see Chapter 1) and coauthors [2]. And then of course the computer hardware itself is based on perfect single crystal silicon with which one deliberately alters the electronic properties with doping to create n- and p-type semiconductors. Probably the most famous discovery of the last century was the structure of DNA, which was based on helical X-ray diffraction data from the fibres of DNA, a fantastic piece of basic science including the basis of understanding heredity. That discovery has proved so important in modern-day genetics and DNA fingerprinting in forensic science. So, all of these things, right the way from natural curiosity all the way through to the utility of science, I think appeal to people. These sorts of examples led to formal approval by the United Nations and UNESCO, of the International Year of Crystallography in 2014, which was brought about through the efforts of the International Union of Crystallography and the Moroccan Crystallographic Association, led by Prof. Dr Sine Larsen and by Prof. Abdelmalek Thalal, respectively, to make the proposal in the first place. This has been a marvellous event for our field and for communicating about our science. There are many marvellous things that happened through the IYCr; one I especially liked was the crystal structure a day through the whole year on Twitter, led by Dr Helen Maynard-Casely and our Australian colleagues and now a calendar at the IYCr legacy website. I think all readers of the IUCr Newsletter would agree that science is necessary. We should not be complacent however and should take note of Max Perutz's book entitled *Is Science Necessary?* [3]. We do have to take time and care to explain our work. However, to guard against us becoming too utilitarian, in his chapter 'How to become a scientist', he captures the heart of a scientist I think when he writes: 'Without delight and wonder at the works of Nature you might as well join Scotland Yard instead'.

REFERENCES

1. Authier, A. (2013). *The Early Days of X-Ray Crystallography*, Oxford University Press, Oxford, UK.
2. Bugg, C.E., Carson, W.M. and Montgomery, J.A. (1993). Drugs by design: Structure based design, an innovative approach to developing drugs has recently spawned many promising therapeutic agents, including several now in human trials for treating AIDS, Cancer and other diseases. *Sci. Am.*, December, 92–98.
3. Perutz, M. (1989). *Is Science Necessary? Essays on Science and Scientists*, Published in 1991 by Oxford University Press, Oxford, UK.

CHAPTER 3

When Did I Begin to Serve the International Community of Science?

I FOUND IT THRILLING TO be invited by the International Union of Crystallography (IUCr) during my career to serve the worldwide crystallographic community in various ways. Why was this? Through my life I did admire other people in the way they served society and one such was Dame Mary Warnock, noted for chairing UK Government committees such as on *in vitro* fertilisation (IVF) [1], and who was recognised with the international Dan David Prize for Biothics in 2018. Since then, IVF births following her committee's recommendations have totalled 4 million. My influences on the IUCr community, although broad, I felt can always be put in the context of the greater achievements of others such as Mary Warnock. The roles that I was successively invited by the IUCr Executive Committees of their time were:

- Founding Chairman of the Commission on Synchrotron Radiation (1990–1996);

- Joint Main Editor of the IUCr *Journal of Synchrotron Radiation* (1994–1999);

- Editor-in-Chief of IUCr Journals (1996–2005);

- IUCr Representative to the International Council for Scientific and Technical Information (ICSTI) (2005–2014);

- Co-Editor of the IUCr *Journal of Applied Crystallography* (2005–2014);

- IUCr Representative to the Committee on Data of the International Science Council (CODATA) (2011–2020);

- Chair of the IUCr Diffraction Data Deposition Working Group (DDDWG) (2011–2017) and then Chair of its successor, the IUCr Committee on Data (2017–2020).

Finally, I have served approaching 30 years as a Member of the IUCr/OUP Book Series Committee and finally as its Chair (2017–2020).

Where did it all start? My first lecture at an IUCr World Congress was in 1981 in Ottawa when I spoke about the software improvement 'round robin' evaluation project I had initiated as a beamline scientist at the UK's Synchrotron Radiation Source [2]. My logic was simple: namely that since I and the SRS were investing all this effort in developing a beamline for protein crystallography, the software needed to be as good as possible. I was then invited to be a member of the IUCr 1984 Programme Committee (PC) for the Congress to be held in Hamburg; the PC met in Pittsburgh where we stayed in the University of Pittsburgh Faculty Club. Professor John White, a very learned neutrons scientist based at the Australian National University in Canberra, and I were in charge of the instrumentation and methods part of the Congress program.

The first Commission task I had was then as follows, and I quote the report of the IUCr Executive Committee for 1989 [3]:

> Ad interim Commission on Synchrotron Radiation. At its meeting in July 1989, the IUCr Executive Committee decided to set up this Commission ad interim until the Bordeaux General Assembly (in 1990). It invited J. R. Helliwell to serve as Chairman, and he readily accepted. A draft set of terms of reference was drawn up by him, after wide consultation with members of the relevant scientific community, and was then approved by the Executive Committee. The agreed draft terms are being published in a variety of crystallographic newsletters. The goals of the Commission, briefly, are to provide a focus of organization and information for various aspects of diffraction experiments at synchrotron-radiation sources worldwide, and so facilitate access to appropriate instruments, to maintain and improve standards and to organize meetings. Proposals for the membership of the commission were made and were accepted by the Executive Committee. All the proposed members agreed to serve on the ad interim Commission, which has now started its work.

After an initial period and evaluation of our work, the *Ad interim* Commission was formally ratified into a full-fledged Commission. This role introduced me fair and square to the organisational workings of the IUCr, a clearly careful and rational organisation. A major role of the Commission on Synchrotron Radiation was to compile beamline specifications to aid users in planning their experiments. An initiative that we strongly supported and actively joined in was the proposal for a *Journal of Synchrotron Radiation* by Dr Samar Hasnain, a colleague at the

Synchrotron Radiation Source in Daresbury Laboratory, UK. This proposal was successful, the new title was established within the IUCr family of journals, and I joined Samar as a Main Editor, as did Dr Hiromichi Kamitsubo, then Director of the SPRing8 synchrotron in Japan. A major aspect of the decision taken by the IUCr Executive Committee of the time, President André Authier, was that the journal would not only cover X-ray crystallography, but it would also include all regions of the electromagnetic spectrum and accept papers on diffraction, spectroscopy and imaging. The significance of this was that one science community had the chance to take the initiative for all fields, and the IUCr took on that opportunity and responsibility. The *Journal of Synchrotron Radiation* has become very successful.

I joined the Journals Commission in 1990 as a Co-editor for *Acta Crystallographica*. This editorial work dictated my Saturday afternoons as I unpacked each submitted article's envelope that would arrive during the previous week, all communication then being by letter. It was an amazing feeling being invited by IUCr President Philip Coppens and the Executive Committee of 1995 to become the Editor-in-Chief of *Acta Crystallographica* (the longest established of the IUCr journals, now published in several sections). As the number of the journal editors was more than one hundred, this led me to chair meetings of these science experts from all over world. Another formal task was to attend the IUCr Finance Committee meetings. The first of these was in March of 1996 to overlap with the outgoing Editor-in-Chief Charlie Bugg from the USA, and who was a mentor in my first year as a doctoral student. I attended nine more of these finance meetings, one each year at about the same time, and so became acquainted with the Hotel Kong Arthur in Copenhagen in late winter, where we always stayed.

During my tenure as Editor-in-Chief, the existing validation procedures for chemical crystallography submissions, already world-leading, were developed further. This now also included our new journal, *Acta Cryst E*, a successful online rapid-publication journal whose entire existence arose from the ability to automatically validate supporting data. I tried to encourage the biological structural community to embrace a similar paradigm of data validation alongside the submitted article. But I failed however in my proposal to the Open Meeting of the Commission on Biological Macromolecules at the 2002 IUCr Congress held in Geneva, where there were nearly 100 persons attending from around the world.

I have continued this cause to the present day! Nevertheless in my interviews for the British Library science audio project [4], I said that I regard that moment at the 2002 Congress as the biggest failure of my career.

Following on from my role as Editor-in-Chief, I was asked to be IUCr Representative to the ICSTI. I met an incredibly diverse group of people from science, as well as science publishers and national libraries worldwide. The working scientists present at these meetings were the representatives of the International Union of Pure and Applied Physics (IUPAP), Professor Sir Roger Elliott, a condensed matter theoretical physicist at Oxford University, and the International Union of Pure and Applied Chemistry (IUPAC), Dr Wendy Warr, a chemical informatics specialist. At my first meeting, Roger asked me 'what are we doing here?', his point being that the proceedings were dominated by non-scientists. My reply was 'to provide science experience as an anchor to keep the others in touch with realities'. Roger and I got on extremely well. Wherever the meetings were held, he always knew a good-value high-quality restaurant. He invited me to St John's College, Oxford 'Rawlinson dinner', a very splendid affair. In return, it seemed to me quite appropriate that I invited him to be my guest at a Manchester United home match, which we duly went to. I was very sad when he passed away in April 2018. I completed my three triennia as IUCr Representative to ICSTI in 2014.

I started my role as IUCr Representative to CODATA in 2011. CODATA is an organisation that was established in 1966 by the

International Council for Scientific Unions as its Committee on Data (CODATA). Around the same time, in 1969, an intergovernmental agreement set up the ICSTI. CODATA and ICSTI complemented each other's activities. Steadily, electronic publishing began to dominate, and so the publishing and information interests of ICSTI considerably overlapped those of CODATA. I suggested, therefore, in my reports to the IUCr Executive Committee that both organisations should be merged, but which has never happened. My first CODATA task was to attend the biennial CODATA Congress and General Assembly in Taipei. I accompanied the outgoing CODATA Representative Brian McMahon, Research and Development Officer at IUCr Chester. Brian knew so much about CODATA and had been a major contributor to the development of the IUCr's data standardisation project, the Crystallographic Information Framework (*CIF*). It was in Taipei that one of the other delegates, Dr Mary Zborowski, a trained physicist from Canada, showed me how to use Twitter, which has become my main source of science and professional news; I am very grateful to Mary for this.

During this time, I was invited by IUCr President Professor Dr Sine Larsen at the Madrid Congress in 2011 to Chair the IUCr DDDWG to look into whether archiving of crystallographic data, already common practice in many journals and databases, should be extended to the primary experimental data, termed the raw diffraction data images? The synchrotron radiation facilities, she said, were looking closely at this, including the ESRF in particular where she was Science Director. The DDDWG's work ran for two triennia and our final report was published at the IUCr Forum website in 2017 [5]. The DDDWG's work, although mainly by email was supplemented by important workshops held in Bergen, Rovinj, and New Orleans, ahead of major international crystallography meetings in those places. We also reported to the IUCr General Assembly in Montreal in 2014. Our DDDWG recommendation that the primary 'raw' diffraction data 'should' be archived was softened by the IUCr Executive Committee into 'encouraged'. One can assume that such a radical transition should not be too sudden. The implications of our final report for various branches of crystallography have continued to be considered by the various IUCr Commissions since its publication. The DDDWG was absorbed into the Standing Committee on Data, a committee whose terms of reference were considerably broader than the DDDWG, which I have chaired since 2017.

More quietly perhaps than the various roles above, I have served on the IUCr/OUP Book Series Committee since 1993. My first encounter with the Book Series was in 1990 when Dr John Robertson from Leeds University approached me about authoring a book within the Series. I said, 'John I am terribly sorry but I have already made a proposal to Cambridge University Press (CUP) to write a monograph on synchrotron radiation' [6]. I had picked CUP because I had always liked the look and feel, as well as the content, of Michael Woolfson's *X-ray Crystallography* [7] and Dr Stephen and Professor Henry Lipson's *Optical Physics* [8], which CUP had published. Anyway the IUCr Executive Committee invited me to join the Book Series Committee, which I duly did. I served under several different Chairs, each with a different mode of working. One encouraged discussion amongst members of proposals and another didn't. None of the Chairs gave a glimpse of what happened after we had submitted our reports to them. This represents, in retrospect, a lack of transparency to committee members that sits uncomfortably with contemporary trends towards openness in review. I only got to know the details of the working agreement and procedure between IUCr and OUP when I was invited in 2017 to become Chair after Professor Dr Davide Viterbo had passed away whilst in office. A key activity I undertook as a Member of the Book Series was to propose and, after its acceptance, write a monograph within the Book Series, which I duly did with colleagues [9].

Overall, not only have I been glad to serve such a worthwhile scientific community-driven organisation as the IUCr, but these roles have also greatly enriched my science experience and provided opportunities for me to make many friends. Why do I think I was repeatedly invited to take on another major IUCr role? Maybe it was because no one else was willing; ah, but, that surely cannot have been the case! Anyway, I always undertook the work seriously, not easy when I had a very busy job either as a scientific civil servant or university professor. Suffice to say that I think my decisions, which were many, were always evidence based. Simply put, I did my best with these various roles.

REFERENCES

1. Warnock, M. (2002). *A Memoir: People and Places*, Duck Editions. First paperback edition (27 March 2002). See also: https://www.dandavidprize.org/laureates/2018/present-bioethics/baroness-mary-warnock.
2. Hellwell, J.R. (2016). Overhead transparencies at IUCr Ottawa 1981 re Protein Crystal Oscillation Film Data Processing: A comparative study & SRS 7.2 Abstract 1981 Ottawa. My overhead transparencies are archived here https://zenodo.org/record/166325.
3. Report of the Executive Committee (1992). *Fifteenth General Assembly and International Congress of Crystallography*, Bordeaux, France, 19–28 July 1990; *Acta Cryst. A* **46**, 871–896.
4. Audio interviews on my scientific career with the British Library and held in their archives. The weblink is: https://sounds.bl.uk/Oral-history/Science/021M-C1379X0122XX-0001V0 See episode 13 22 minutes 30 seconds in for my comments on refereeing of article, with data, with validation report.
5. Helliwell, J.R., McMahon, B., Androulakis, S., Szebenyi, M., Kroon-Batenburg, L.M.J., Terwilliger, T.C., Westbrook, J. and Weckert, E. (2017). Final report of the IUCr Diffraction Data Deposition Working Group. https://www.iucr.org/resources/data/dddwg/final-report.
6. Helliwell, J.R. (1992). *Macromolecular Crystallography with Synchrotron Radiation*, Cambridge University Press, Cambridge, UK, paperback (2005).
7. Woolfson, M.M. (1971). *An Introduction to X-Ray Crystallography*, Cambridge University Press, Cambridge, UK, 2nd Edition 1997.
8. Lipson, S. and Lipson, H. (1969). *Optical Physics*, Cambridge University Press, Cambridge, UK.
9. Chayen, N. Helliwell, J.R. and Snell, E.H. (2010). *Macromolecular Crystallisation and Crystal Perfection*, Oxford University Press, Oxford UK.

CHAPTER 4

Where Did My Scientific Career Lead Me?

Not long after my formal retirement from my academic position in August 2012, I was awarded the prestigious international prizes; the Patterson Award of the American Crystallographic Association in 2014 and the Max Perutz Award of the European Crystallographic Association in 2015.[1] This led to the interview below by the IUCr, in which I was invited to reflect on my career and its particular highlights.

Reproduced with permission of the International Union of Crystallography and is available at https://www.iucr.org/people/crystallographers/eighth-perutz-prize.

---

[1] This interview was conducted by Jonathan Agbenyega of the IUCr and is reproduced here with permission of the International Union of Crystallography. The original text is available at https://www.iucr.org/people/crystallographers/eighth-perutz-prize.

A POTTED HISTORY OF YOUR CAREER TO DATE?

I have worked in the scientific civil service (at Daresbury Laboratory) and in academia. For the latter, I have been on the faculty in two physics departments (Keele and York Universities) and one chemistry department (Manchester University). My research career started in a department of zoology at Oxford University. As well as my DPhil in molecular biophysics, I have a Doctor of Science in Physics, and for the last 26 years, I have been a Professor of Structural Chemistry, now Emeritus. My research and teaching has involved students and postdocs from many countries in the world; so as a researcher and educator, one gets a very wide vision of the nations and peoples of the world. Overall I would describe myself as a trained crystallographer and an interdisciplinary scientist and educator; I have published research results from biological crystallography, chemical crystallography, physical crystallography and on instrumentation and experimental techniques; I have taught a wide variety of undergraduate courses and supervised postgraduate students trained in chemistry, physics and biological sciences.

HOW DID YOU GET INTO THE FIELD OF CRYSTALLOGRAPHY?

I first got interested when I was an undergraduate. I did physics at York University and I took option courses in X-ray crystallography and in Biophysics. These were given by really leading researchers and educators, Professor Michael Woolfson FRS who taught the X-ray crystallography and Dr Peter Main (Honorary Member of the British Crystallographic Association) who taught the Biophysics. So, it went from there really. That was the genesis. I went on to do a DPhil in protein crystallography in Oxford with Dr Margaret Adams. I might have become an astrophysicist. I was keen on that as well but I always enjoyed chemistry at school so it was perhaps quite natural that I moved towards a mixture of physical and chemical sciences in my research. At York, I also found I enjoyed computing, and I was good at maths at school, so that was overall a good mixture to enable me to do crystal structure analysis I think.

WERE THERE ANY MAJOR MILESTONES ALONG THE WAY AND KEY TURNING POINTS?

The major aspects are maybe best measured by when one makes major changes in one's career. The phases I can identify, punctuated by changes as milestones, involved: first, a constant feature, was that I had approximately four decades involved with synchrotron radiation instrumentation and methods development, including the very important services to industry. These were all at the UK's Daresbury SRS and also supporting the European Synchrotron Radiation Facility (ESRF) initiative in numerous ways, first at CERN in Geneva (where the European Synchrotron Radiation Project, ESRP, was initially based before ESRF was built) and then in Grenoble. As well as those I served on various SR Facility Advisory Committees around the world. A second theme was my research in physical, biological and chemical crystallography and various molecular structural studies research programmes. A third theme is that I have taken on some of the biggest tasks one can in the service of the crystallographic community. Fourthly, I have undertaken crystallographic researches with neutrons at the Institut Laue Langevin these last two decades and most recently assisting the European Spallation Neutron Source in Lund. Finally, I take pride in my students and what they have achieved and the major challenges they have taken up.

WHAT HAVE BEEN SOME OF YOUR OWN MEMORABLE AND PERSONAL ACHIEVEMENTS TO DATE?

I have touched upon aspects of this above and I explained more in my eighth Max Perutz Award acceptance lecture at the ECM29 Opening Ceremony in Rovinj, Croatia. I would like to add something about trying to move gender equality in science forward.

My DPhil supervisor was female (Dr Margaret Adams) as was her DPhil supervisor (Prof. Dorothy Hodgkin OM FRS). I didn't personally recognise gender equality as an issue but when I did I tried to do something about it, notably as gender equality champion in my School of Chemistry in Manchester University. I led a working group on behalf of the School towards firstly a Bronze Award and then a Silver Award of achievement from the UK Government's Department of Business, Innovation and Science 'Athena SWAN' scheme (SWAN=Science Women's Academic Network). The major issue is of course the widespread inequality that women experience in making their way in science. A perhaps less highlighted issue is that the natural talent that we train to a very high standard as graduates in chemistry and physics and so on, respectively around 40% and 20 % females, leads to only around 5% of women becoming professors. This is a massive discrepancy. We have to understand each stage of the academic career progression path and do much better at assisting women at each of these stages. Otherwise we will continue to see this huge loss of trained talent. That is what the Athena SWAN scheme is about, a marvellous initiative of our UK Government, and I am glad to have played a role in trying to help. Crystallography as a field has a better than average performance and for instance we have seen three female IUCr Presidents; that is good but much remains to be done.

CHAPTER 5

When Did I Become a Writer of Popular Science Books?

L IKE MANY A WORKING scientist, I have written a significant number of scholarly articles, monographs and reports for fellow professionals. But in 2016, I made a proposal to a publisher, CRC Press, of the need for a concise handbook on the skills needed for a life in science. The success of that book, judged by the various very favourable reviews of it, has led to a series of further essays and reflections, of which the current volume is a member. My publisher introduced these books to prospective readers through the interviews reproduced below.

Q&A WITH JOHN R. HELLIWELL ON THE PUBLICATION OF HIS BOOK *SKILLS FOR A SCIENTIFIC LIFE*

What Led You to Writing It?

Following my role as Senior Mentor for New Academics for the School of Chemistry at the University of Manchester and my long-time Manchester Gold experience as a mentor, I decided there was a useful role for a book on *Skills for a Scientific Life*. My conviction was reinforced by my long-time Scientific Civil Service experience based at Daresbury Laboratory, not least as CCLRC Director of Synchrotron Radiation Science, where I took a close interest (and had to!) in annual appraisal for a staff member's objectives setting and career development. Indeed, in the Scientific Civil Service, skills training and development was compulsory. As a Director I attended *standards of management excellence training*, experience which I have treasured ever since. The University of Manchester also has a staff development unit and runs excellent training courses, various of which I attended.

Can You Describe Your Book in One Sentence?
My book provides my career-long insights and case studies into the skills needed by a scientist as a researcher and educator.

Who Would Be Interested in Reading Your Book?
My book is aimed at several groups of people:

- Scientists engaged in research, development and discovery.

- People considering starting 'Science' as a career.

- School pupils and students deciding whether to study science subjects versus arts and humanities subjects, and likewise their advisors and their parents.

My book describes what science is like, including the successes, as well as the challenges, and yes, some trials and tribulations. This book will be valuable to people at all stages of their science careers.

Are There Any Relevant World Issues That
Your Book Relates to at the Moment?
My book invites the reader to consider how best to make their own global impacts in science, consider how to change the

current organisation of science and aim to make a contribution to world peace and to sustainability. The societal impacts that a scientist can realise can be achieved via their lectures and press releases to the public, and which are both satisfying and needed also by the funding agencies that scientists receive their research grants from. This links with my other recent book, *Perspectives in Crystallography,* which offers a threefold look into the past, present and future of crystal structure analysis. Crystallography is one of the most multidisciplinary sciences, with roots in fields as varied as mathematics, physics, chemistry, biology, materials science, computation and earth and planetary science. The structural knowledge gained from crystallography has been instrumental in acquiring new levels of understanding in numerous scientific areas. This book resonates with the 2014 United Nations and UNESCO International Year of Crystallography, a celebration of its achievements and importance, undertaken with the International Union of Crystallography (IUCr). Crystallography in both its organisation within IUCr and in its discoveries offers major contributions to sustainability including within the United Nations' Millennium Development Goals.

Tell Us an Unusual Fact about Yourself and Your Teaching

I have an unusually wide experience teaching physicists, chemists and biological scientists at undergraduate and postgraduate research levels.

Q&A WITH JOHN R. HELLIWELL ON THE PUBLICATION OF HIS BOOK *THE WHYS OF A SCIENTIFIC LIFE*

What Do You Want Your Audience to Take Away from the Book?

First, let's define the audiences I am aiming at. The understanding of why scientists do what they do should and does interest several audiences namely the public, politicians, schoolchildren and their parents and last but not least scientists themselves should consider these topics.

What Inspired You to Put This Book Together?

CRC Press's Hilary LaFoe responded very positively to my ideas for a new book. Specifically, she suggested that CRC Press had a new book series on Global Science Education. I thought the concept for this was excellent. I also thought that an excellent place to start would be *The Whys of a Scientific Life*, the book title in my mind. This title arose because my previous book on *Skills for a Scientific Life* had chapters nearly all defined as "How to....". So, a complementary title would now be "Why". Nearly every one of my chapter titles is "Why...".

Why Is This Book Relevant to the Present Day?

My book is relevant to people of all ages. Also, it is relevant to scientists today where they are more and more exhorted to explain their research and its wider impacts.

How Do You Think Your Field Is Evolving Today?

Science policy is evolving because Governments and Funding Agencies seek to direct the research that is done to be more and more obviously for the good of Society. Such a policy seems laudable, and in some ways is, but some of the most impactful discoveries in science were not immediately obvious as having applications. I explore these perspectives in my book.

What Are the Main Developments in Research That You Are Seeing in Your Subject Area of Expertise?

There are many and I do give some practical examples of my own in my new book which illustrate the general discussion I am elaborating on. I also refer everyone to my recent book *Perspectives in Crystallography*, also published by CRC Press in 2015, where you can see a detailed explanation, which answers your question.

What Makes Your Book Stand Out from Its Competitors?

I genuinely believe my book has no competitors with its combination of modern-day policy matters, careers relevance and philosophy of science in the context of modern developments especially the hugely improved scope for primary, i.e. raw data archiving arriving finally at objectivity.

Is There One Piece of Research Included in the Book Which Surprised You or Challenged Your Previous Understanding of the Topic?

The pace of research and development taken as a whole outstrips an individual's vision and imagination, including mine. It is this cooperative enterprise overall that science is. I mean 'overall' to allow for the fact that individuals can compete fiercely and not cooperatively.

What Did You Enjoy about Writing This Book?

I especially enjoyed researching each of the "Why" chapter topics and sometimes being moved to quote some policy matter announcements.

What Advice Would You Give to an Aspiring Researcher in Your Field?

As the back cover to *The Whys of a Scientific Life* elaborates, scientists are driven by their curiosity but they do of course work within a complex environment. That is a challenging mix to try and navigate. So, I hope my book is of help to scientists too.

Who Was/Is Your Role model? Who Inspired You to Pursue a Career In...?

I found the research work of Max Perutz on using X-ray crystallography to understand the workings of the oxygen transport protein haemoglobin inspiring. This was during my biophysics option course at the University of York when I did my physics degree. I didn't realise then that I would spend so much of my career working on instrumentation and methods developments, in effect that X-ray crystallography was far from perfect. Its scope is now transformed out of all recognition. My ideas, and frustration with the state of the art of instrumentation in 1975, led me to have discussions with Dorothy Hodgkin, also a Nobel Prize winner like Max Perutz, and which also profoundly changed my research career.

Q&A WITH JOHN R. HELLIWELL ON THE PUBLICATION OF HIS BOOK *THE WHATS OF A SCIENTIFIC LIFE*

What Do You Want Your Audience to Take Away from the Book?

Again, let's define the audiences I am aiming at with this third book in the *Scientific Life* Series. I undertook to survey the landscape of what is the scientific life because it is necessary for the public and schoolchildren to gain that overview. But also scientists are so often entrenched in their particular discipline that they are rarely if ever exposed to an overview.

What Inspired You to Put This Book Together?

Again CRC Press's Hilary LaFoe responded very positively to my ideas for my new book. Specifically, she suggested that the CRC Press new book series on Global Science Education was the best venue for it. Since the scope of my new book was so broad, it was going to be challenging to meet the short format style of books in the series. By opting for a personal experience as evidence approach, I believe I achieved my aim. By this, I mean I necessarily side stepped trying to write a historical-based tome, which would have had to be 1000 pages or more!

Why Is This Book Relevant to the Present Day?

The landscape of science is very broad and its applications extensive. Society recognises the importance of science and needs to be able to read a relatively short account of the current landscape of science. I hope also to inspire schoolchildren by including descriptions of how I did in my school and university science.

How Do You Think Science Is Evolving Today?

The merger of the International Council for Science with the International Social Sciences Council to form the International Science Council is a huge step but a necessary one which I have supported.

What Are the Main Developments in Research That You Are Seeing in Your Subject Area of Expertise?

There are many and I do give some practical examples of my own in my new book which illustrate the general discussion I am elaborating on.

What Makes Your Book Stand Out from Its Competitors?

Again I genuinely believe my book has no competitors because of its breadth without it being a heavy tome. The Global Science Education Book Series is an ideal vehicle for it.

My three *Scientific Life* books were given book launches hosted by my publisher Hilary LaFoe of CRC Press and held at Blackwell's Bookshop, Manchester Oxford Road branch, with introductory remarks by Dr Diana Leitch (MBE, FRSC, Deputy University Librarian, University of Manchester, and Chair of Trustees of the Catalyst Science Discovery Centre and Museum, Widnes, UK). Here is what she said about my book *The Whats of a Scientific Life*:

> A lot of thought about what science is, and experience about what the scientific life involves, has gone into this book by John Helliwell. In fact John has become a figure head for his work in physics and chemistry and also delving into biology. His book is within the Global Science Education Book Series and which, close to my heart, is the important science education role that the Catalyst Science Discovery Centre and Museum in

Widnes has and where I am the Chair of Trustees of the Charitable Trust that manages it. I warmly welcome this new book, not least in which John highlights enzyme catalysis and the International Year of the Periodic Table 2019, it even includes a photograph of Mendeleev's visit to Manchester.

CHAPTER 6

When Did I Become Passionate about World Peace?[1]

Without peace there is no stability for progress.

I BECAME DEEPLY INTERESTED IN peace in the world as a DPhil student learning enzyme crystallography at Oxford University with Dr Margaret Adams. We were based in an overall group headed by Dorothy Hodgkin and her people, Guy and Eleanor Dodson and research visitors. I learnt that Dorothy was to become President of Pugwash (1976–1988), an organisation whose main objective was 'the elimination of all weapons of mass destruction and of war as a social institution to settle international disputes'. Dorothy was also International Union of Crystallography (IUCr) President in this period (1981–1984). Coming from a

---

[1] Originally published in the *Newsletter* of the International Union of Crystallography (IUCr) at https://www.iucr.org/news/newsletter/volume-27/number-4/what-we-can-do-towards-world-peace. Reproduced by permission of the IUCr and the Newsletter Editor, Emeritus Professor Mike Glazer.

family where my father and grandfather had fought in World Wars, I had a simple view that defence, not aggression, is what is necessary. Pugwash deepened but did not resolve my thinking away from this view. Later, by chance, I came across Kathleen Lonsdale's short book *Is Peace Possible?* [1]. My thinking was swept forward by her book and its careful analysis summed up in the Christian saying 'turn the other cheek'. But, I think this can be so difficult when I thought of the scale of offensive provocations, or threats of, that we witness or know of from the past. Then I discovered Linus Pauling's book *No More War* [2], which focused on nuclear weapons. Since the first use of nuclear weapons, the peace that has followed since 1945 is surmised to be due to the knowledge that a nuclear war would be of such magnitude that it really could not be allowed to happen.

In this somewhat random walk in deepening my thinking about these matters another step unexpectedly came in the *Times Higher Education Supplement* of 14 November 2019 where I saw a review of the book *World Peace (And How We Can Achieve It)* by Alex J. Bellamy [3]. I bought an e-copy of the book which I have read carefully. It tackles the same issues afresh. That nuclear weapons are the mainstay of the modern peace since 1945 is aired in Bellamy's book on page 94.

Should we crystallographers take an interest in all this? More to the point what can we do towards peaceful ends? It isn't in our science, *but* it is in our history because of Lonsdale, Hodgkin and Pauling. So I think we should take a direct interest. I have written about this before. Several years ago when I saw that the American Crystallographic Association's Transactions Symposium in 2015 was on *Crystallography and Sustainability*, I submitted an abstract. In my subsequent article [4], I blended our historical crystallography leaders' activities, which I mention above, with aspects of my most recent research on green (structural) chemistry applications, namely striving for clinical outcomes with less chemicals. In my research for the talk at the ACA 2015, I was struck by the absence of striving for peace as one of the United Nations eight, at the time, sustainability development goals (SDGs), a fact which I found baffling.

So I looked to this new book by A. J. Bellamy to learn more about these issues and what one as an individual can do to try to help. Bellamy has extensive experience and is very accomplished in the field of peace studies. There are several points to be highlighted, although the whole book is a very interesting treatment of the topic. It also provides a very long historical look back and scrutinises cave paintings as evidence of peace or war-like activities of our ancestors, and much more besides. Suffice to say, he finally suggests improvements in our modern procedures for peace upheld as best it can by the United Nations.

More specific to us as crystallographers, I immediately checked if Bellamy's book cited Kathleen Lonsdale's book or Linus Pauling's book. But he does not. I also wondered if scientists' roles are 'only' the shortening of wars, e.g. by the sound-ranging initiatives of W. L. Bragg in the First World War [5]. But there was no mention of W. L. Bragg either; maybe that is a too specific contribution for me to hope for. But Bellamy's book has introduced me to J. D. Bernal's 1958 book *World without War* [6], although Bellamy gives Bernal the briefest of mentions. Bernal at page 17 [6] is uncompromising in his analysis of the nuclear deterrent: since 'each side could destroy the other and between them destroy the whole world why not accept the nuclear stalemate and dismantle the whole apparatus?' Bernal's book goes on to describe his vision that 'Only in a world without war is the possibility to satisfy human needs.' Indeed it is still true today, as Bernal pointed out then, that the expenditure on violence per year is $14.8 trillion, i.e. 12.4% of the world's GDP [7]. Imagine what humankind could do if it would agree to peaceful objectives and thereby, on a huge scale, tackle poverty and the widespread lack of opportunities in so many places.

Anyway the historical context of 50–100 years ago is one thing but what about now? The current era is best described in Bellamy's Chapter 8 entitled "Towards World Peace". The UN Charter is described and to a degree is dissected. There is clearly a specific problem of the power of a veto or 'threatened vetos' of the members of the UN Security Council. This does seem a 'fundamental flaw in the international security architecture' (p. 185, [3]). I also agree that the 'challenge isn't to reinvent the institutional architecture but to make it work better'. On the last page of Bellamy's book (p. 213, [3]) is a call to 'everyone to assume some responsibility towards world peace'. The final quotation goes to the UN Secretary General Ban Ki-Moon in 2016:

> Peace is not an accident. Peace is not a gift. Peace is something we must all work for, every day in every country.

I had not come across this before. It is a very important statement. No matter where we are, or when, there is an urgency to realising peace.

CERN in Geneva was instigated not only to find out the new fundamentals of physics but also as a peace project, bringing scientists of many nations together in a peaceful enterprise. This concept connects firmly with what our scientific community is doing now in facilitating the Middle East synchrotron in Jordan[2] and helping with the African Light Source project[3]. The IUCr itself is directly active as a promoter of these with its *Light Sources for Africa, the Americas, Asia and the Middle East (LAAMP) Project*,[4] which IUCr is participating in along with the International Union of Pure and Applied Physics under the auspices of the International Science Council. Our IUCr Congresses also require free circulation of scientists[5] another vital ingredient for peace. We crystallographers are striving to improve ourselves with respect to gender balance within all our activities; Bellamy in multiple places in his book emphasises the importance of

---

[2] https://www.sesame.org.jo/.
[3] https://www.africanlightsource.org/.
[4] https://laamp.iucr.org/.
[5] https://council.science/what-we-do/freedoms-and-responsibilities-of-scientists/freedom-of-movement-and-association/.

gender equality for societies striving for peace. As crystallographers, we can then all take many direct roles as our contributions towards peace in the world.

REFERENCES

1. Lonsdale, K. (1957). *Is Peace Possible?* Penguin, Harmondsworth, Middlesex, UK.
2. Pauling, L. (1958). *No more War!* Dodd, Mead and Company, New York.
3. Bellamy, A.J. (2019). *World Peace (and How We Can Achieve It)*, Oxford University Press, Oxford, UK.
4. Helliwell, J.R. (2015). Crystallography and sustainability. In: *Transactions Symposium Crystallography for Sustainability*, pp. 8-19, American Crystallographic Association, Buffalo, NY. https://www.amercrystalassn.org/assets/volume45.pdf.
5. Van der Kloot, W. (2005). Lawrence Bragg's role in the development of sound-ranging in World War I, *Notes Rec. R. Soc.* **59**, 273–284.
6. Bernal, J.D. (1958). World without War. *USA National Library of Congress.* Available online here: https://babel.hathitrust.org/cgi/pt?id=mdp.39015065741962&view=1up&seq=18.
7. Institute for Economics and Peace (2018). *Global Peace Index 2018: Measuring Peace in a Complex World.* Sydney, Australia. ISBN: 978-0-6483048-0-7; www.economicsandpeace.org.

II

Where to Look for Your Inspiration? Select Your Role Models; Here Are My Role Models in Science

The following sections are only a small selection of the great scientists who have inspired me in many ways. Two others I should certainly mention are my excellent DPhil supervisor Dr Margaret Adams, Tutor in Inorganic Chemistry at Somerville College, Oxford University, but the DPhil is more of a one-to-one working relationship. Also, Professor Dorothy Hodgkin OM FRS (Nobel Prize in Chemistry 1964) influenced me directly during that time in Oxford. Dorothy's life has of course been well documented both within and outside the crystallography community, and I would particularly recommend the lovely, detailed, memoir by Guy Dodson [1]. Two more inspiring scientists for me are Dr Olga Kennard OBE FRS, founder of the Cambridge Structure Database in 1965 and joint founder of the Protein Data Bank in 1971, and Professor Eleanor Dodson FRS for her many contributions to the collaborative computational protein crystallography ('CCP4') project. Olga presented an inspiring to all of us opening lecture for the CSD 50th Celebration in 2015 [2]. She was into her nineties! Eleanor, into her eighties, is frequently answering queries posted on the CCP4 community bulletin board! Also, Professor Michael Rossmann from Purdue University who sadly passed away in 2019 was an inspiring scientist presenting break-through studies in biomolecular science into his eighties.

REFERENCES

1. Dodson, G.G. (2002). Dorothy Mary Crowfoot Hodgkin, O.M.12 May 191029 July 1994. *Biogr. Mems Fell. R. Soc. Lond.* **48**, 179–219.
2. Kennard, O. The Cambridge Crystal Structure Database at 50: Origins, influences and directions. https://www.youtube.com/watch?v=HkR7_uxvU8Q.

CHAPTER 7

When I Was at the Physics Department at York University, UK; My Memories of Professor Michael Woolfson FRS

I WAS AN UNDERGRADUATE IN the physics department from 1971 to 1974. The university was founded in 1963 initially without science subjects. The physics department's first cohort of undergraduates arrived in 1965 and so the courses and teaching laboratories when I was being taught were all well prepared and tested. In 1971, we were a group of 32 physics undergraduates, made up of 30 young men and two young women. By young I just mean that all of us had entered the university at about 18 years of age. On the general culture of the department at that time I would mention the following. York University had been my third choice. My first two choices, Durham and Bristol University physics departments, rejected me. Later my pride was somewhat restored when I learnt that not being from a private school probably explained my rejections by them. Certainly through my whole experience at York, I never felt any prejudice against my state educated school background.

Michael Woolfson was a remote figure to me until our third year when he taught me three courses; X-ray Crystallography, Astrophysics and Advanced Quantum Mechanics. To us he was Professor of Theoretical Physics, a figure of authority obviously. But he was indeed a marvellous lecturer providing clear and rigorous presentation and diction, and deserving of our respect.

In general, there were advantages to being in a small class. First, it was quite straightforward to speak to a lecturer straight after their lecture if we had a question. Michael was quite approachable. Second, each teaching laboratory experiment was done by oneself, which I still think is a major advantage so as to learn self-reliance. I recall that Michael had designed one of those experiments but sadly I cannot remember which. Clearly Michael had diverse interests in physics. In third year, we undergraduates would pick a two-term project; Michael offered a project on formation of the planets. He also offered an essay exam question on the biophysics of vision. In my year, he did not offer a crystallography third-year project. That was offered by Dr Peter Main, which that year was on the crystal structure of urea. I decided on a transmission electron microscope project on determining the Burgers' vectors of dislocations in thin metal films supervised by Professor Martin Prutton. It was during this project period that I learnt that Michael and Professor Prutton commuted together as Michael lived in Leeds and Professor Prutton in Boston Spa. Somehow through this project period, presumably from Professor Prutton, I learnt firstly that Michael was Jewish and the Leeds Jewish community was quite large and active. I later learnt that perhaps Michael did not want to live in York because:

> of one of the worst anti-Semitic massacres of the Middle Ages which took place in York in 1190. The city's entire Jewish community was trapped by an angry mob inside the tower of York Castle. Many members of the community chose to commit suicide rather than be murdered or forcibly baptised by the attackers.[1]

As an undergraduate, I launched the York University Astronomical Society with the help of Dr Rod Greenhow, who looked after the physics department's telescope. It was here that I first saw Saturn's beautiful rings. Michael accepted our Society's invitation to speak on the formation of the planets, which we were enthralled by. I also organised a coach trip to Scarborough to visit the planetarium there. Michael and his wife Margaret came along with us. They were probably mystified by my enthusiasm for the semi-mythical starry light show there. Anyway in retrospect, I decided that it showed that the York Astronomical Society was to be encouraged.

Michael also recruited me back to the department in 1985 as a lecturer when he was head of department. My wife and I were very happy in York. Madeleine started work as the chemistry department's crystallographer through the helpful auspices of Professor Guy Dodson. I stayed until the end of 1988 before moving to the University of Manchester as Professor of Structural Chemistry; how could I turn down a chair! During those years in the York physics department, I taught the X-ray crystallography course that Michael had taught. I also taught the biophysics option course, which I had also taken as an undergraduate when it was taught by Dr Peter Main. This biophysics course in 1974 was the one that had excited me about doing a doctorate in

---

[1] https://www.english-heritage.org.uk/visit/places/cliffords-tower-york/history-and-stories/massacre-of-the-jews/.

protein crystallography, especially learning about Max Perutz's crystallographic studies on the movement of the iron atom in haemoglobin on oxygenation. With that biophysics course as the inspiration and having taken Michael's X-ray crystallography course, with his famous textbook on X-ray crystallography [1] in hand, I felt confident to do a DPhil in Oxford on enzyme crystallography. This move to Oxford meant that I had also to reverse my decision to accept an offer from Professor Prutton of a PhD in surface physics at York.

In this period when I was lecturer in the physics department, Michael and I won a research grant from the Science and Engineering Research Council (SERC) on direct methods development for protein crystallography. It was a big award and we recruited the husband and wife team from Calcutta, Drs Alok and Monika Mukherjee. Monika worked with Michael and Alok with me. One of these projects working with Peter Main led to using X-ray wavelength optimised anomalous differences with the direct-methods program *MULTAN* to locate the metal atoms in a protein crystal [2]. These were diffraction data I had measured from crystals of the iron containing cytochrome and the manganese and calcium containing pea lectin at Daresbury SRS 7.2, which I had designed to be a fully tuneable instrument. This work using *MULTAN* built on Keith Wilson's initiative in locating heavy atoms using isomorphous differences with MULTAN [3], which I had promptly applied in my DPhil [4].

Michael was very generous to me in assisting my research career development. An enquiry to him from Mrs Dora Gomez de Anderez from Venezuela was transferred over to me, and Dora worked with me for her PhD. For Dr Hao Quan's request to come from the Beijing Institute of Physics and undertake crystallography research, Michael both supported the Royal Society Queen Elizabeth Fellowship application and let Hao Quan do his research with me. On the occasion of Michael's 65th birthday, I dedicated an article to him [5] entitled *Towards the*

Measurement of Ideal Data for Macromolecular Crystallography Using Synchrotron Sources:

> This paper is dedicated to Professor M. M. Woolfson FRS, University of York, on the occasion of his 65th birthday, who would remark to the author 'Our new direct method works with ideal protein crystal data but not with real data'.

While I was on the academic staff at York, Michael was President of the British Crystallographic Association (BCA). The annual conference of the BCA took place in York in 1986. I assisted Michael in the practicalities. This experience stood me in good stead when I chaired the 1993 conference in Manchester.

In October 1985, the Nobel Prize in Chemistry was awarded to Jerome Karle and Herbert Hauptmann 'for their outstanding achievements in the development of direct methods for the determination of crystal structures'.[2] Obviously this was one of Michael's core research interests [6], with huge achievements of his own. I wondered why Michael had not been included, since up to three people could share the Prize. The matter was however obviously complicated for the Nobel Committee, it seemed to me, as Dr Isabella Karle and Dr David Sayre had strong credentials too. I remember when Michael was awarded the American Crystallographic Association's (ACA) A. L. Patterson Award in 1990 in New Orleans. I attended that conference of the ACA and vividly recall Michael and Herbert Hauptman sitting together in convivial conversation. That made a big impression on me as it showed science as an endeavour that enjoys 'friendly' competitiveness. In 1994, I gave a lecture at the Hauptman Woodward Medical Research Institute in Buffalo. I met Herbert Hauptman there of course. The Institute had some pamphlets, one of which was on the origins of direct methods and the dream of Herbert

---

[2] https://www.nobelprize.org/prizes/chemistry/1985/summary/.

Haupman as a mathematician to develop and apply them for the benefits of medicine. In the text was a mention of a young British physicist, Michael Woolfson, who had written that direct methods would not work. From this, I learnt the importance of being careful to never say that something wouldn't work.

Michael Woolfson made a big contribution to the development of synchrotron radiation policy in the UK. In November 1991, the SERC commissioned him to chair its second six-yearly review of synchrotron radiation science:

> A multidisciplinary panel, chaired by Professor M. M. Woolfson FRS, has been established to undertake this review. The task of this panel is to assess and report on the UK community's needs for access to synchrotron radiation facilities in the next century and, if new capital programmes were indicated, the form they might take.[3]

In April 1993, the 'Woolfson Report', as it became known, was published. The report

> recognised the quality, volume and diversity of science being carried out at synchrotron radiation facilities and put forward a number of recommendations for continued support. Among these was the major recommendation that a new medium-energy X-ray source be constructed to replace the UK's existing Synchrotron Radiation Source (SRS) at Daresbury, Cheshire.

The SRS had commenced its user programme in 1980 or so. The proposed new medium energy source sought to still complement the high-energy synchrotron radiation source, the European Synchrotron Radiation Facility (ESRF) in Grenoble, in which the UK was a founding country partner. Workshops to detail the proposed ESRF commenced in the 1980s, with significant input

---

[3] https://publications.parliament.uk/pa/cm199900/cmselect/cmsctech/82/9121503. htm.

from Daresbury Laboratory on the machine and experimental aspects. The third source recommended by the Woolfson Report [7], a low-energy synchrotron, has never been funded, although a detailed proposal for 'Sinbad', as it was called, was developed at Daresbury Laboratory. The scientific and technical case for the new medium-energy synchrotron was accepted in 1994 by the UK Research Councils. In 1997, the Council for the Central Laboratories of the Research Councils (CCLRC) – the body then responsible for operating the SRS – completed an initial feasibility and design study for the new facility.

Despite general agreement throughout the scientific community that the UK needed a new synchrotron radiation facility in order to maintain its position as a major world player in this field, problems over financing delayed action for several years. These problems were finally resolved, including a significant financial commitment by The Wellcome Trust. The siting of the new synchrotron was decided in late 1999 to be at the Rutherford Appleton Laboratory, Oxfordshire and it was named the Diamond Light Source. Diamond produced its first user beam in 2007. As the largest single scientific infrastructure project for many decades, it is fair to say that the Woolfson Report of 1993 had a truly major impact on science in the UK and for many decades to come.

REFERENCES

1. Woolfson, M.M. (1997). *An Introduction to X-Ray Crystallography*, Cambridge University Press, Cambridge, UK. The first edition was published in 1969.
2. Mukherjee, A., Helliwell, J.R. and Main, P. (1989). The use of MULTAN to locate the positions of anomalous scatterers. *Acta Cryst.* **A45**, 715–718.
3. Wilson, K.S. (1978). The application of MULTAN to the analysis of isomorphous derivatives in protein crystallography. *Acta Cryst.* **B34**, 1599–1608.
4. Adams, M.J., Helliwell, J.R. and Bugg, C.E. (1977). Low resolution structure of 6-phosphogluconate dehydrogenase. *J. Mol. Biol.* **112**, 183–197.
5. Helliwell, J.R., Ealick, S., Doing, P., Irving, T. and Szebenyi, M. (1993). Towards the measurement of ideal data for macromolecular crystallography using synchrotron sources. *Acta Cryst.* **D49**, 120–128.

6. Woolfson, M.M. (1987). Direct methods: From birth to maturity. *Acta Cryst.* **A43**, 593–612.

7. Woolfson, M.M., Engineering and Physical Sciences Research Council (1994). *Review of Synchrotron Radiation Science: A Summary of the Report of the Panel Chair*, Engineering and Physical Sciences Research Council, Swindon, UK, 27 p. ISBN: 1870669894. A copy of the Report with its appendices and annexes, including a separate Audit Committee Report, is available from the Chadwick Library, Daresbury Laboratory.

CHAPTER 8

When a Researcher; My Collaborations with Professor Durward Cruickshank FRS

DURWARD CRUICKSHANK WAS A great collaborator for me from 1982, when he formally retired from the University of Manchester Institute of Science and Technology (UMIST).[1] He was an eminent scientist having established his reputation on his pioneering studies of the precision of crystal structures through his whole career. Our research was principally on the foundations of the Laue method, with Professor Keith Moffat of Cornell University, USA, which led to new, and we believe major,

---

[1] The article by Beagley, B. and Helliwell, J.R. (2018). *Biographical Memoirs of Fellows of the Royal Society*, **65**, 73–87, provides the complete description of Durward's life and career.

developments in the use of synchrotron radiation and neutrons in crystal structure analysis.

An important aside is for me to give some of Keith's career core details. Keith graduated in physics from Edinburgh University, including being taught by Professor Peter Higgs of 'Higgs boson' fame and who was awarded the Nobel Prize in Physics in 2013. Keith then went to do his PhD with Max Perutz, Nobel Prize in Chemistry 1962, at the Medical Research Council Laboratory of Molecular Biology in Cambridge. Keith moved to Cornell University and finally to the University of Chicago for the remainder of his career. In both Cornell and Chicago, he was the prime mover behind establishing beamlines for macromolecular crystallography at first the Cornell Synchrotron and then the Advanced Photon Source in Chicago.

Durward's academic career started in engineering studying at Loughborough College, as it was known then, where he obtained an external London University degree in 1944. Subsequently, he studied mathematics at Cambridge, 'learning at the feet of Bondi, Hoyle, Boys and Dirac', as Durward put it to me, being awarded successively a BA (1949), an MA (1954) and finally a DSc (1961).

His scientific curiosity was fashioned at school during his war-time senior years when St Lawrence College, Ramsgate, was evacuated to Courteenhall, Northamptonshire. With limited teaching on hand, the boys largely taught themselves. Of the six or so science students leaving between 1942 and 1943, three – including Durward – went on to be elected Fellows of the Royal Society.

Durward quite possibly was the only Chemistry Professor never to pass a chemistry exam. Not that he even took one, for his academic background was in engineering and mathematics, and his war-time interlude work was in the Special Operations Executive working on mini-submarines. When his daughter showed interest in chemistry at school, he would arrive home with all sorts of chemicals from the UMIST labs, where he was Professor, for her chemistry experiments.

Between 1944 and 1946, he worked for the Admiralty on Naval Operational Research matters including underwater submersibles, specifically the *Welfreighter*. This was a four-man midget submarine designed to land agents and supplies in occupied territories for assisting Resistance Movements. Durward was based first at The Frythe, a country house in rural Hertfordshire which had been commandeered in August 1939 by British Military Intelligence. Then, during the *Second World War*, it became a secret British *Special Operations Executive* factory, known as Station IX, making commando equipment. Secret research included military vehicles and equipment, explosives and technical sabotage, camouflage, biological and chemical warfare. In the grounds of The Frythe, small cabins and barracks functioned as laboratories and workshops. Durward remarked to me that in a matter of 6 weeks, he learnt a great deal about probability and statistics that was of practical use.

In our many discussions, Durward maintained that he owed his career as an X-ray crystallographer to Sir Gordon Cox FRS, and he introduced me to Cox at a British Crystallographic Association annual conference. Unfortunately, I cannot exactly remember which year that was. Cox had invited Durward to Leeds where, as Professor of Chemistry, he had established a thriving chemical crystallography unit. In 1946, Durward joined Cox's group as a temporary research assistant. Initially, he was assigned to dismantling old hospital X-ray and former military equipment from which Cox asked him to construct a structure-factor calculator. 'This worked – just', he wrote. He studied the mathematics of crystal structure refinement. His mathematical insight from his days as an engineer enabled him to see ways to develop this area. His focus was on accurate crystal structures [1], as mentioned in my introduction, which continued throughout his whole career.

In September 1950, he was appointed Lecturer in Mathematical Chemistry in the University of Leeds, but before returning to Leeds, he attended the Cambridge Summer School on Programme Design for Automatic Digital Computing Machines. This was amongst the world's first summer schools on electronic computing. This School introduced Durward to the principles of computer programming, machine order codes and binary arithmetic. He also learnt of the progress of the Manchester University Electronic Computer Project and that The Ferranti Mark I computer would be available to outside users in 1952.

For the Manchester Ferranti computer, Durward wrote programming code in assembler. With this he undertook his careful and rigorous calculations of atomic thermal vibrations. With his first research student Farid Ahmed, from Canada, he sorted out conflicting earlier results on the carboxylic group in oxalic acid dihydrate [1]. They obtained bond distance estimates for the C=O double bond of 1.187 ± 0.022 Å and for the C-O(H) single bond of 1.285 ± 0.012 Å   [2]. These particular bonds featured strongly in his final article, on proteins, in 2007 [3].

How to sum up such a research career spanning 60 years? Durward was an eminent scientist, whose mathematical abilities transformed the precision of the molecular structures determined in three dimensions by X-ray crystal structure analysis. He had a direct influence on over one million chemical crystal structures, the number currently determined and held in the Cambridge Structure Database alone. And, oh yes, he had a profound and direct influence on me as a researcher spanning 25 years, and I still think of him since he passed away in 2007. In the article [4], I made the following dedication to Durward:

> Dedicated to Professor Durward W. J. Cruickshank, FRS, Emeritus Professor of Chemistry, The University of Manchester, on the occasion of his 82nd Birthday, 7 March 2006. Durward is an inspiration to us all as a scientist with many fine contributions spanning nearly

60 years, as well as a friend, mentor and colleague to several of us for many years.

REFERENCES

1. Cox, E.G. and Cruickshank, D.W.J. (1948). The accuracy of electron-density maps in X-ray structure analysis. *Acta Cryst.* **1**, 92–93.
2. Ahmed, F.R. and Cruickshank, D.W.J. (1953). A refinement of the crystal structure analyses of oxalic acid dihydrate. *Acta Cryst.* **6**, 385–392.
3. Ahmed, H.U., Blakeley, M.P., Cianci, M., Cruickshank, D.W.J., Hubbard, J.A. and Helliwell, J.R. (2007). The determination of protonation states in proteins. *Acta Cryst.* **D63**, 906–922.
4. Cianci, M., Helliwell, J.R., Helliwell, M., Kaucic, V., Logar, N.Z., Mali, G. and Tusar, N.N. (2005). Anomalous scattering in structural chemistry and biology. *Crystallogr. Rev.* **11**(4), 245–335.

CHAPTER 9

When I Discovered the Efforts for Peace by Esteemed Crystallographer Dame Kathleen Lonsdale FRS

DAME KATHLEEN LONSDALE WAS one of the first female Fellows of the Royal Society and the first woman professor at University College London.[1] She was a key person in British and international science as indicated by two examples of her accomplishments: she was President of the International Union

---

[1] Adapted from Helliwell, J.R. (2012). *Crystallography Reviews* **18**, 33–93 and references therein. Reprinted by permission of the publisher (Taylor & Francis Ltd, http://www. tandfonline.com).

of Crystallography and was the first female President of the British Association for the Advancement of Science (BAAS). She was also a leading pacifist of her time. She trained as a physicist and became a professor of chemistry.

One of her famous crystallographic analyses was that of hexamethylbenzene [1], analyses which showed that the bond distances around the aromatic benzene ring were equal and not alternating short and long distances of single and double bonds. Thus, it is a resonance molecular structure. She had a strong interest in determining the mobility of atoms from crystal structure analyses, and thereby inferring details of reactions in crystals and the crystalline state. She remained in essence a crystal physicist, and apart from W. H. Bragg (Nobel Prize in Physics 1915 jointly with his son Lawrence Bragg), the most important influence on her scientific outlook was probably Michael Faraday. Working in Faraday's room at the Royal Institution, she read his notebooks and absorbed his approach to experimentation. Michael Faraday (1791–1867) discovered electromagnetic induction, diamagnetism and electrolysis.

Her obituary in *The Times*, in noting that Kathleen Lonsdale, was the 'First woman President of the British Association for the Advancement of Science', stated that '(she) aimed darts at a variety of targets: the sale of arms, the narrowness of many scientists and (the) responsibility of scientists as a whole for the use made of their discoveries'. In the same role, she delivered a lecture at the Leeds meeting of the BAAS in 4 September 1967 on *Physics and Ageing.*

Her most profound influence on me was that she articulated her views on peace so well. In her letter to the Minister of Labour and National Services, Whitehall SW1 on 29 May 1942 she wrote:

Sir,

...

I am absolutely opposed to the principle of *universal* compulsory registration and conscription *for war purposes*. As a member of the Society of Friends (Quakers) I believe war to be the wrong method of resisting aggression or any other form of evil… In refusing, therefore, to comply with the regulation or to take advantage of possible exemption on conscientious or other grounds, and in being prepared to take the consequences, I am acting in accordance with what I believe to be my highest duty. I rather wish I did not.

Yours sincerely,
Kathleen Lonsdale (DSc London).

She served 1 month in Holloway Prison in 1943 as a result.

She expressed her pacifist views in detail in her book: *Is Peace Possible?* (Figure 9.1). The front cover of the book has the headline: *A Quaker scientist discusses problems of peace, freedom, and justice in an era of expanding world population and technical development.* She researched her topics carefully; she had clearly studied, and quotes from, the UN Charter, as well as, e.g. giving a detailed summary of the stages and landmarks in the Israeli–Palestinian conflict, and of the role of Britain and other countries in that conflict (her Chapter 9 of [2]). The book also contains her strongly held religious beliefs. She sums up her arguments and logic developed in the book as follows (p. 127 of [2]):

> …a life of non-violence is essentially one of deep spiritual out-reach to the good in other men and of belief that, even if there is no response, even if we appear to fail, goodness will in the end prevail. *Yea, though I walk through the valley of the shadow of death I shall fear no evil, for Thou art with me.*

As an example of her passion for both peace work and science, she describes in her book [2] the hopes of the time after the Second World War, namely for a

> peace settlement that would end all war. It might take time… Meanwhile my work was fun. I often ran the last

FIGURE 9.1 The front cover of my copy of Kathleen Lonsdale's book *Is Peace Possible?*

few yards to the laboratory. Later on I took my mathematical calculations with me to the nursing-homes where my babies were born; it was exciting to find out new facts.

REFERENCES

1. Lonsdale, K. (1929). The structure of the benzene ring in $C_6(CH_3)_6$. *Proc. R. Soc. Lond. A* **123**, 494–515.
2. Lonsdale, K. (1957). *Is Peace Possible?* Penguin, Harmondsworth, Middlesex, UK.

CHAPTER 10

When and Where I Met a Giant of Indian Science; Professor M. Vijayan, Who Became President of the Indian National Science Academy

PROFESSOR M. VIJAYAN, A National Academy of Sciences India Platinum Jubilee Senior Scientist based at the Molecular Biophysics Unit Indian Institute of Science, Bangalore, India, has retired and shared with us this remarkable autobiography of his life and work.[1] It is a wonderful memoir. I feel grateful to him for writing of all his experiences and contributions to science which are so vast, nationally and internationally.

I know Vijayan quite well, having met him on numerous occasions at International Union of Crystallography events and also at International Union of Pure and Applied Biophysics (IUPAB). Eventually he became the President of the Indian National Science Academy. In 1999 at the New Delhi IUPAB, I led the UK Delegation. Vijayan was there of course. I remember how Professor Dr Richard Ernst (Nobel Prize in Physics 1991) hugely overran his Congress Plenary lecture time slot, firmly disrupting the whole day! This incident is not mentioned at all in Vijayan's memoir and reminded me of the Indian characteristic of politeness. So it is a measure of the frankness in which he displays his anger about British colonialism's impact in India (p. 189):

> The British also promoted English education in the country with the objective of utilizing the services of Indians at lower levels of administration and commerce. As an unintended consequence, a segment of Indians was exposed to modern science and western liberal ideas.

Vijayan's youth was inspired very much by politics and communism in particular before his interests in science overtook his politics. Nevertheless the memoir describes a remarkable amount of committee work at numerous levels that he undertook, thus combining his community science vision with his science research passion.

Vijayan and I had similar backgrounds, both of us with a degree in undergraduate physics and then specialisation in

---

[1] This section is based on Helliwell, J.R. (2020). A review of *A Life among Men, Women and Molecules Memoirs of an Indian Scientist* (2020). Edited and Coordinated by A. K. Singhvi. Published by the Indian National Science Academy Bahadur Shah Zafar Marg, New Delhi, India. *Crystallography Reviews*. DOI: 10.1080/0889311X.2020.1788005. Reprinted by permission of the publisher (Taylor & Francis Ltd, http://www.tandfonline.com).

molecular biophysics from an early stage, mine in my DPhil and his in his first postdoctoral post with Dorothy Hodgkin in Oxford. His two periods in the same laboratory as I were such that our respective years in Oxford meant we just missed each other. I am struck how with our common origin in physics, I strived to retain my physics, whilst being a professor of chemistry, and so I call myself an amateur whole biologist, whilst a practising structural molecular biologist. His book documents his full conversion to molecular biology and his commitment to establishing macromolecular crystallography in India. He provides lovely detail of his work on the insulin project in Oxford with Dorothy Hodgkin and the lifelong friendships that ensued. There are many lovely pictures such as with Guy and Eleanor Dodson over the decades that Vijayan shares with us.

His molecular studies of ensembles of static crystal structures, such as with hydration as variable, gave him structural dynamics insights. He led crystallographic studies of many different combinations of peptides, thus mimicking the origins of life of the different biological molecules meeting and reacting in clay minerals to form the first polypeptides.

The evolution of macromolecular crystallography (MX) in India and the technology instrumentation needs described in the book are particularly interesting to me. I read how his first area detector basically stilled the immediate need for synchrotron access for Indian crystallographers for rapid data MX collection, and his success with multiple isomorphous replacement (MIR) phasing of his new protein crystal structures stilled the need for multiple wavelength anomalous dispersion (MAD) synchrotron capabilities, as it became known. India did not proceed as expeditiously as Vijayan hoped to realise its own national synchrotron X-ray source. Vijayan interestingly summed up the situation on page 159:

> I believe that we would have been well on the way to
> a new synchrotron facility, but for clashes involving

king-sized individual and institutional egos. I do not know when and how the efforts can be revived. If and when it is done, developments in free-electron laser technology and cryo-electron microscopy also need to be taken into account. Despite the disappointment in relation to an Indian facility, our sustained campaign was not fruitless. Assured access to ESRF (in Grenoble) and Elettra (in Trieste) and the commissioning of Indus-2 (in Indore, Madhya Pradesh) with the active involvement of the user community, were major achievements.

Vijayan steadily rose up through the hierarchies of Indian science. He was asked to receive on behalf of G. N. Ramachandran the IUCr Ewald Prize in 1999 at the IUCr Glasgow Congress, which Vijayan describes with lovely humility.

I have very warm memories of India including my several visits to the Molecular Biophysics Unit in Bangalore, and my visit to Calcutta for the Professor K. Banerjee Centennial Symposium in 2000, and also IUPAB 1999 New Delhi, and the IUCr 2017 in Hyderabad. Vijayan himself was a person one always wished to meet on one's travels. This is a book of experiences and insights to treasure and learn from.

III

Career Development

CHAPTER 11

When Should I Look to My Employer to Facilitate My Career Development?

DIFFERENT COUNTRIES APPROACH THIS in different ways. The UK approach I know in most detail. I have lived through it these last nearly 50 years, since I entered research in 1974 as a doctoral student at Oxford University. Over time, I was successively a postdoctoral researcher, a scientific civil servant (in the USA, for example, this would be called a government scientist), through to being a professor running a laboratory responsible for permanent staff, fixed-term postdoctoral researchers and doctoral students. Then I was a director of a large facility and later a senior mentor for new academics. Researcher career development is very important. Indeed, is one's biggest and best contribution the research publications that one has produced or the training of new researchers? On both counts, one does one's best. The research environment is intricate. I never doubted at any stage of my research career that any one of my employers behaved other than responsibly. That was actually obviously not true as the documented failures for equality and diversity subsequently show over the decades. When I joined a large academic department in 1989 as a professor, there were no female academics, out of nearly 50 and yet ~40% of our undergraduates were female. This is but one example that there was relatively little, compared to now, checking for signs of discrimination.

So, what is the structure of the researcher career development environment in 2020? In the UK, it is guided by a government-commissioned Concordat, launched in 2008 and reviewed, by a different government appointed body, in 2018: (https://www.ukri.org/files/skills/concordatreviewreport-jun2018-pdf/ and https://www.ukri.org/files/skills/csg-concordat-review-response-september-2018-pdf/). A core point in this Concordat is how does a researcher develop independently, if they should wish to? (Recommendation 4, see below).

The Independent Review Report of the Concordat to support the career development of researchers included the following:

> Recommendation 4: There should be increased support for researcher independence, including autonomy in their own career development, and the freedom to innovate. A revised Concordat should address the tension between Principal Investigators (PIs) and postdoctoral independence, setting out clearly the obligations for both groups. There should be increased emphasis and support, by both funders and employers, for uptake of researchers' ten days training allowance. Development of researcher independence should be supported through allocated time within grants. 20% of a researcher's time should be allowed for developing independent research and skills.

The Concordat Strategy Group responded to that very important Recommendation 4 of the Review on a researcher's career development as follows:

> The CSG agrees with the principle that there should be increased support for researchers to develop their individual research identity and professional competencies and looks forward to engaging with the community on this positive recommendation. The development needs of individual researchers may vary in intensity and nature over time and may be realised differently depending on individual needs and disciplinary contexts. We also recognise that researchers should be aware of the wide range of employment opportunities open to them within and beyond academia and understand how they can develop and apply their knowledge, experience and competencies to be competitive and successful in diverse occupations. The CSG welcomes discussions on how research staff can develop their capabilities at an appropriate level and welcome examples of existing good practice across the sector.

I note that this response carefully avoided agreeing with the 20% 'researcher's own research and skills development' time. A specific employee's input on Recommendation 4, which to my mind correctly stated:

> The concordat makes no provision that employers should guarantee researchers any personal career development time, nor does it specify or suggest what reasonable time should be allowed. I consider this should reasonably be around 20% of contract time (as is often specified in academic and teaching contracts). In my job description I am allocated 2.5% of my time for personal / career development. I have colleagues with no written guarantee of personal / career development time in their contract.

The way the principles are currently written allows institutions (or individual PIs) to require that most personal
development, e.g. publications, training and attending
conferences, will take place in a researcher's own time.
This can be discriminatory against those with health
issues, disabilities and caring responsibilities (often
women).

As evidenced by that employee's input, and on such details, the
future of a scientific research enterprise overall depends. The UK
has stated an aim to considerably expand its R&D endeavours
with considerably more funding, which is good. But the wisdom
of Max Perutz as Director of the world famous Laboratory of
Molecular Biology in Cambridge is very pertinent here namely
'that I left people to do what happened to interest them'. The early
career researcher has to be given the proper time and resource
to develop as well as realising the objectives of the principal
investigator's project. If the project is 3 years, then fund it for the
20% more that is the early career person's due, as specified in the
Review Report Recommendation 4 above.

CHAPTER 12

Where Do You Want to Get to? Career Road Maps

Are career road maps even feasible? To some degree I think they are; as the saying goes 'chance favours the prepared mind'.

The jobs.ac.uk website offers helpful practical advice about developing an academic career. One of their excellent guides[1] starts from the transition from undergraduate to postgraduate and onwards.

A common challenge is where two people who share their lives together are both making their individual careers, often both in very competitive workplaces. Some employers do not even allow partners to both work there. In road maps, planning for such a double career, working in cities naturally offers more flexibility. There is a wider choice of employers in cities, and there are also

---

[1] https://www.jobs.ac.uk/media/pdf/careers/resources/a-practical-guide-to-planning-an-academic-or-research-career.pdf.

likely to be larger employers with more staff and therefore more vacancies.

Early on in my career, a senior colleague helpfully mentioned to me a way to help understand the direction where you want to head to. He said 'ask yourself who do you admire?' Therefore that can be a way to understand your career goal wishes and thereby assist you in choosing the steps on your career path.

I would also advise that I would not worry unduly about how quickly one reaches a goal. As Buddha is quoted as saying 'There is no path to happiness: happiness is the path.' So, being reasonably happy in your daily work mix is a good guide to whether your current career path suits you. What does 'reasonably happy' mean? As my mother used to say to me 'no job is perfect'.

In mid-career, a feeling can develop that one is stuck at a career ceiling. It is quite likely that an employer will have a mentoring scheme for staff. The University of Manchester has its Manchester Gold scheme. I served as a mentor on this scheme for about 10 years. My mentees were a mixture of early career and mid-career academics. The career ceiling challenge can be tackled and my role as mentor, the critical friend, was to help the mentee understand themselves and their particular work environment better. A first tool in that process is an analysis of what a person spends their time on. Is a disproportionate amount of time spent on particular tasks? How can a better balance of time spent be realised? This may expose problematic situations in the work environment such as others in their workplace creating situations leading to firefighting. Fires (i.e. emergency situations) must be dealt with, but a second step is to put in place ways to avoid that situation happening again. So, what is positive for the mentee can become beneficial for the workplace as well. This is a win-win and likely lead to the mentee being better valued in that workplace. Some workplace environments, and the managers and or senior managers, are not always receptive to such initiatives. That, one can then fairly safely conclude, is then an indicator to quietly look elsewhere to continue one's career.

IV

Specific Examples Across the Disciplines of Science: The When and Where in Basic Science

In physics, the constancy of its laws, everywhere in the universe and forever, is taken as true. The assertion is tested by measuring at ever better precision the fundamental constants of physics. Thus far, the constancy of the laws of physics does hold true. For chemistry, biology and materials science, there are some caveats, but the laws of physics are taken as true once more. For chemistry though, reactions take place at different speeds by raising the temperature or including a catalyst. For biology, we study life and that in turn depends on environmental conditions such as a favourable temperature and pressure, over a certain range. But different organisms are more resilient than others. If we travel yet further in academic discipline to the social sciences, we see a great deal of measurement such as of opinions of populations and assessment of values in societies. Variations over time within the same country or region are self-evident as they are between different parts of the world. Even in one's scientific life, the fellow scientists one meets vary a lot such that their character needs to be regarded carefully, to ascertain their ethics, before collaborations ensue (see Chapters 15 and 31 of my book *Skills for a Scientific Life*). When and where in basic science is then a vast topic, but let's examine a few examples from the physical sciences in the next few chapters.

CHAPTER 13

When But Not Where or Where But Not When? The Non-classical World of Quantum Physics for the Electron and What It Means for Chemistry

*W*HEN AND *WHERE* PLAYS a big role in basic science. The quantum world does not allow an electron's position to be known precisely where and precisely when together. That is very odd. If someone says, meet me at say 6 pm but not precisely where you would rightly be cross! In the quantum world, the best that we can do is to provide a probability distribution for where an electron is. But that still allows a deep understanding of the chemical reactivity of the elements in the Periodic Table.

Nevertheless, Albert Einstein was not happy that an electron could only probably be in a given place; famously he said that 'God does not play dice with the world'. Whilst much has been focussed on the God aspect of this sentence, the probability aspect, the rolling of a dice, is surely the real focus. So, what experimental science supports the reality of the quantum world? If you look at an old-fashioned sodium street lamp, it is yellow. In the physics laboratory, one can look at the light emitted by an electrically excited sodium gas with a prism spectrometer. I first did this experiment in my school physics laboratory. A prism is what Isaac Newton used to show that white light has the continuous colours of the spectrum. Instead, the pure sodium gas has specific spectral lines, i.e. lines of a specific energy or wavelength. How can that be? An initial idea for the structure of an atom due to J. J. Thomson was his 'Christmas pudding' model. Sir Joseph John Thomson, Order of Merit and President of The Royal Society, was a British physicist and Nobel Laureate in Physics in 1906. He suggested that the electrons were held away from the positively charged nucleus because without the pudding, the negatively charged moving electrons would spiral towards the nucleus. This would involve continuous changes of colours of the emitted light from those electrons. Instead, the quantum theory treated the electron itself as a wave and where it had standing wave probability solutions to the quantum theory equation of Erwin Schrödinger, Nobel Prize in Physics 1933, such that the atom would be in a stable state, called the ground state. By exciting the sodium gas, the electron of a sodium atom leaves its ground state taking on the extra energy and has an excited, but still stable, state. The excited atom could then relax back to the ground state and emit specific energy light, i.e. a yellow colour.

So, the when and where of an electron's position and its 'movement' means it has to be treated mathematically by a probability distribution with the mathematics of the Schrödinger equation. This is quite abstract but firmly planted in the foundation of experimental physics of the prism spectrometer. A good question to ask is what about a proton? Yes, it is governed by the same mathematics, but being nearly two thousand times the mass of an electron (1×10^{-30} kg), it is confined to have a local space unlike that of an electron. It can be considered as being the centre 'point' of the nucleus in a hydrogen atom, the simplest element comprising one electron and one proton.

How do the electrons determine the chemistry, i.e. the reactivity of the different chemical elements? As one adds more and more electrons to make a new chemical element, and the corresponding number of protons along with an increasing number of neutrons to stabilise the nucleus of an atom, the electron probability distributions do interpenetrate. In some cases, an element is not quite at stability. It wants to meet another element that has a spare electron that is rather loosely held. Sodium chloride, common salt, is one such. The sodium can quite readily give up its 'outer electron' to the chlorine to complete its outer shell of electrons to acquire a better stability. This is a purely ionic interaction, where an electron is given by one element to the other. In another example, electrons can be shared so as to realise stability, which is called a covalent bond. Two atoms, of the same or different elements, can form covalent bonds. The whole basis of the reactivity of elements, and the great variety of reactivities of all the elements, is the science of chemistry. Some elements have already the optimal number of electrons for stability and so they do not react with other elements. They are inert.

The year 2019 was declared by the United Nations, who approve such things, as the International Year of the Periodic Table (IYPT) of the elements. In my local town of Stockport, various decorative frogs appeared in our shopping centre with different chemical elements and their chemical symbols painted on them.

FIGURE 13.1 Great interest in the International Year of the Periodic table (IYPT) at one of the frogs in Stockport town centre, 12 August 2019. Two directions of view are shown of this frog.

These attracted a great deal of interest from children and their parents. I asked a passer-by if they would please take a photo of me next to one of the IYPT Stockport frogs (Figure 13.1). He said sure, but why? I explained the IYPT and that I was a Professor of Chemistry. He was really interested to know more about chemistry. I didn't give him the full explanation of the quantum world above but did give a tourist-level basic description of how many elements there were and so on as well as explaining the properties of a couple of the featured elements on that Stockport frog. I hope that he would be interested to learn more.

CHAPTER 14

When Again and Where Again? Frequency and Repetition in Physics and Mathematics

W HEN WE SAY 'WHEN' is something going to happen it can be once only. But repeated events are commonplace. We talk of the frequency. This also applies when we say 'where'. In a street, we see repeated street lamps. We say something like the street lamps repeat every 50 m; that is 'spatial frequency'. In mathematical language, because time and frequency are clearly related, they are called conjugate variables. The same is true of position and spatial frequency, which are also conjugate variables one to the other. Jean-Baptiste Joseph Fourier (1768–1830), who was a French mathematician and physicist born in Auxerre, introduced the mathematical procedure, a transform, between such cases of conjugate variables. So useful did his transform become it became known as the Fourier transform. He also was the first to discern the greenhouse warming of the Earth, over and above the heating of the Earth from direct sunlight. I described this discovery by Fourier in Chapter 16 on climate change in my previous book *The Whats of a Scientific Life* [1].

The Fourier transform is incredibly useful. It allows us to understand what are the important, i.e. most prevalent frequencies in either time or position space. When you decide to buy a high-fidelity sound system you are looking for electronic equipment such as the amplifier and the loudspeakers to faithfully convey the full range of frequencies of sounds in a piece of music. The lowest frequencies, the bass notes, particularly require quite bulky loudspeakers. The highest frequency sounds in the music are most demanding of the amplifier. You can compare two different pieces of sound equipment by taking the Fourier transform of each one playing the same piece of music. The best equipment, likely to be the more expensive, will be the one that passes as wide a band of frequencies as possible. We call that band of frequencies a spectrum. The word 'spectrum' is the same word as used for a prism and passing white light through it, showing the colour spectrum. The effectiveness of the sound equipment is called its 'frequency response'.

The same sort of Fourier transform analyses can be done with an image. By calculating a two-dimensional Fourier transform of the two-dimensional image, it will reveal the predominant spatial frequencies in the image. In our simple one-dimensional example of the street with its many street lights, each one placed 50 m from the other, the Fourier transform is very simple. It is a single spatial frequency. If the houses are spaced 75 m apart, then the Fourier transform would show up both the spatial frequency for the street lamps, 'every 50 metres', and the spatial frequency for the houses, 'every 75 metres'.

REFERENCE

1. Helliwell, J.R. (2020). *The Whats of a Scientific Life*. CRC Press Taylor & Francis, Boca Raton, FL.

CHAPTER 15

When and Where to Invest in a New Megaproject? The Proposed UK X-ray Laser

For nearly a year now, there has been an ongoing exercise by the UK Science and Technology Facilities Council (STFC) to[1]

> develop the science case for a potential UK X-ray free-electron laser (XFEL). We are seeking input to the process from across the scientific community. STFC is supporting this activity, on behalf of UKRI, with Jon Marangos (Imperial College) the project lead and John

[1] Helliwell, J.R. "The UK XFEL Consultation exercise". Reproduced from *Crystallography News* with the permission of the Editor and brought up to date on publication of the full science case.

Collier (Director CLF, the Central Laser Facility) the STFC project champion. A key element will be an assessment of the level of interest within the UK Scientific Community.

To that end, there have been an extensive set of Town Meetings. The overall kick off meeting was held at The Royal Society in July 2019. This has a helpful website with the presentation files of most speakers also available: https://www.clf.stfc.ac.uk/Pages/XFEL-royal-society.aspx.

The website contains presentations from most of the speakers outlining the potential benefits to many fields including chemistry, materials, engineering and structural biology.

The STFC emphasised that high-brightness ultra-fast X-ray pulses from an X-ray free-electron laser (FEL) allow the simultaneous imaging of atomic scale structure, electronic state and dynamics in a material. There is no other X-ray source technology that can do that with femtoseconds time resolution. The unique science opportunities that these machines can open up include: access to very fast structural dynamics; new modes of nanoscopic imaging; access to transient states and the potential to capture rare events.

The UK has considered such a new, FEL, light source on home soil several times before. I recall the Fourth-Generation Light Source (4GLS), Sapphire and the New Light Source.[2] Also the UK joined Euro XFEL in Hamburg, then withdrew, then rejoined. We also had participated in helping with instrument developments at the Linac Coherent Light Source (LCLS) at the Stanford Linear Accelerator Center (SLAC), USA.

But I sense a new mood with respect to this latest consultation exercise. Also, dare one think it, the new UK government is maybe looking for a clear demonstration of a homeland initiative, separate and distinct from our participation in European

---

[2] See https://stfc.ukri.org/about-us/where-we-work/daresbury-laboratory/future-light-sources/.

projects, that it can launch with new money. A UK XFEL could be part of taking the UK up towards 3% of Gross Domestic Product (GDP) spent on scientific research.

The Science Case full draft, gathering up all the steerings that were offered, pro and con, at these various Town Meetings, was published in July 2020.[3] It is a consultative document guiding further discussion through 2020. I am enthusiastic about the possibilities for biomolecular structure determination, both time-resolved and static. I prepared a brief presentation entitled 'What is the structural chemistry of the living organism at its temperature and pressure?'[4] for the Newcastle consultation event. I am very much in accord with STFC's declaration, mentioned above, of new modes of nanoscopic imaging: these can be used for seeing the nanoscopic arrangements in nanotechnology and life sciences, free from radiation damage and adverse effects of sample preparation (e.g. in situ imaging of the function of biomolecular assemblies at their operating temperature).

When and where will this new UK X-ray laser project be built? *When* is unclear as the proposal must first pass through several science approval committees, and if it passes through those, it will be subject to a UK Government gateway approval process. *Where* will it be built? UK policy now is, it seems clear, to place all major scientific facilities at the STFC Rutherford Appleton Laboratory near Oxford. This would apparently require a large quantity of new concrete though, thereby climate change expensive. I know the Daresbury Laboratory site well and it offers a 100 m long experimental hall, plus a further 100 m of car park directly behind it, to accommodate a linear accelerator. Then the existing Northern Institutes Nuclear Accelerator (NINA) (see Chapter 1) ring tunnel made of thick concrete walls would form the inner hall for the UK XFEL experimental hutches.

---

[3] https://stfc.ukri.org/files/uk-xfel-science-case/.

[4] My talk is available at Zenodo https://zenodo.org/record/3565339#.XoNIS2Z7ntQ; I have also expanded on this theme recently in an *Acta Cryst. D* article (http://journals.iucr.org/d/issues/2020/02/00/nw5093/index.html).

I am just making a suggestion that is climate change friendly and will reduce costs with the pre-existing building structures just described. In any case is it wise, even without the climate change issues of a new build, for everything to be in one place?

The physics of synchrotron radiation and FELs is brought together in a recent book by Kim et al. [1], which I reviewed in the *Journal of Synchrotron Radiation* [2]. Time-resolved diffraction [3] has received a great boost down to the nanosecond time range by use of polychromatic Laue diffraction at synchrotrons, complementing use of electrons and neutrons as probes of the structure of matter [3], as well as nuclear magnetic resonance (e.g. [4]). The extension into the femtosecond range by X-ray lasers is a breakthrough for chemical reaction dynamics studies.

REFERENCES

1. Kim, K.-J., Braun, A. and Huang, Z. (2017). *Synchrotron Radiation and Free-Electron Lasers: Principles of Coherent X-Ray Generation*, Cambridge University Press, Cambridge, UK.
2. Helliwell, J.R. (2018). Synchrotron radiation and free-electron lasers: Principles of coherent X-ray generation. *J. Synchrotron Rad.* **25**, 625–626.
3. Helliwell, J.R. and Rentzepis, P.M. (eds) (1997). *Time-Resolved Diffraction*, Oxford University Press, Oxford, UK.
4. Fenwick, R.B., van den Bedem, H., Fraser, J.S. and Wright, P.E. (2014). Integrated description of protein dynamics from room-temperature X-ray crystallography and NMR. *Proc. Nat. Acad. Sci. USA*, **111**, E445–E454.

CHAPTER **16**

When to Move Fast Yet Take Care

Contributing to the Understanding of Coronavirus, COVID-19 in 2020

SURPRISINGLY, WORKING FROM HOME in the UK Government's Lockdown proved to be productive for me. Various of the synchrotron X-ray sources were kept running to allow scientific staff working at the beamlines to collect diffraction data from crystals of the proteins from the COVID-19 virus, also known as SARS-Cov-2.

These diffraction data were shared both for the raw experimental diffraction datasets at several data archives as well as the processed data and protein model atomic coordinates at the Protein Data Bank. I undertook checking of these as did quite a number of other crystallographers around the world. More generally, the preservation of raw diffraction data in new large data archives had become the policy of the International Union of Crystallography.

Secondly, colleagues, including former PhD students, contacted me by email to discuss the various results and the new experiments that they could contribute, such as from the synchrotron facilities, the neutron research centres and the X-ray laser centres, as well as the checking and rechecking the deposited results thus far.

Thirdly, a *Physics Today* writer, David Kramer, contacted me, having been given my name by the Chief Executive of the American Institute of Physics. David wanted to write a news piece on all the work going on around the world on COVID-19. This led to a widely read publication in *Physics Today* [1].

The aim of these crystallography studies was to find a compound that would block the attachment of the virus at the human cell surface receptor or one of its key protein's functions. This, if one such was discovered, was the second line of defence if the vaccination approach to resist the disease, the first line of *innate immunity* defence, failed. Or, both treatments could be used of course. Figure 16.1a–c shows the crystal structure of the main protease of the COVID-19 with a bound inhibitor and the amino acids that it interacts with [2]. This was determined by Chinese scientists very early in the pandemic (Deposited in the Protein Data Bank as code 6m0k on 22 February 2020). Figure 16.1d shows the room temperature crystal structure determined by X-ray and neutron crystallography at the Oak Ridge National Laboratory, USA (Deposited in the Protein Data Bank as code 7jun on 20 August 2020) [3]. The bound water molecules are seen in Figure 16.1d as "boomerangs", and these waters are displaced by the inhibitor on its binding to the protein. The more waters that are displaced the bigger the Gibbs free energy change favouring the binding, a desirable property of a drug. The role of neutrons in biological structural molecular science, in the context of the pandemic is described by Blakeley and Schorber (2020) [4].

Before these scientifically based medicine remedies are adopted, first and foremost is to stop the pandemic situation of

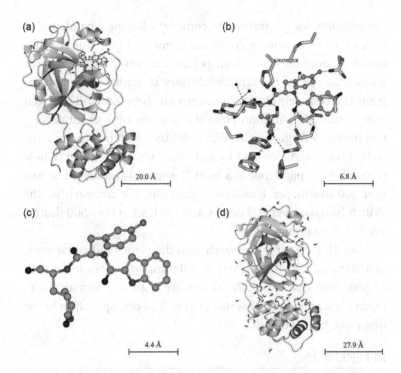

FIGURE 16.1A–C shows the crystal structure of the COVID-19 main protease with a bound inhibitor and the amino acids that it interacts with, determined by Chinese scientists early in the pandemic (Based on Protein Data Bank code 6m0k [2]. Figure 16.1d shows the room temperature crystal structure determined by X-ray and neutron crystallography (Based on Protein Data Bank code 7jun [3]. The bound water molecules are seen in Figure 16.1d as "boomerangs" each with their central oxygen atom bound to two hydrogen atoms. I prepared these figures using CCP4mg [5].

the COVID-19. The ideal situation to be in for a country is to be an island or group of islands. This allows for tight controls on who is let into the country by plane, boat, rail and road. Anyone who is allowed to travel into the country can be escorted to their preferred lockdown address for a 14-day quarantine. Also, it is a relatively much simpler matter, than locking down a whole economy, later risking losing many tens of thousands of lives and wrecking the economy for generations to come by allowing a pandemic to gain a hold on a country. An island nation that took this sensible, basically common sense, strategy has been New Zealand. In such a case, the role of the science advisory committee to a government is a relatively simple one. Thus far, New Zealand has seen four deaths per 1 million population, at my time of submitting this book, 25 August 2020. When did they act?; before the virus could take a hold. For land-locked countries, an exemplar of how it handled the pandemic has been Germany as measured by just over 100 deaths per 1 million population. For comparison: the British Isles has over 600 deaths and the USA at over 500 deaths, per 1 million population.

Overall, for current research into the virus (and consequent activities, such as testing and distributing vaccines), it is worth emphasising that when the stakes are high (i.e. human lives), Open Data and external scrutiny as well as prompt action in the first place help avoid disaster.

REFERENCES

1. Kramer, D. (2020). World's physics instruments turn their focus to COVID-19. *Phys. Today.* https://physicstoday.scitation.org/doi/10.1063/PT.3.4470.
2. Dai, W., Zhang, B., Jiang, X.M., Su, H., Li, J., Zhao, Y., Xie, X., Jin, Z., Peng, J., Liu, F., Li, C., Li, Y., Bai, F., Wang, H., Cheng, X., Cen, X., Hu, S., Yang, X., Wang, J., Liu, X., Xiao, G., Jiang, H., Rao, Z., Zhang, L.K., Xu, Y., Yang, H., Liu, H. (2020). Structure-based design of antiviral drug candidates targeting the SARS-CoV-2 main protease. *Science* 368, 1331–1335.
3. Kneller, D.W., Phillips, G., Weiss, K.L., Pant, S., Zhang, Q., O'Neill, H., Coates, L. and Kovalevsky, A. Unusual zwitterionic catalytic site of SARS-CoV-2 main protease revealed by neutron crystallography. *J Biol Chem.* 2020 Oct 15:jbc.AC120.016154. doi: 10.1074/jbc.AC120.016154. Epub ahead of print. PMID: 33060199.
4. Blakeley, M. P. and Schorber, H. (2020). Neutron sources join the fight against COVID-19 CERN Courier Available online here: https://cerncourier.com/a/neutron-sources-join-the-fight-against-covid-19/
5. McNicholas, S., Potterton, E., Wilson, K. S. and Noble, M. E. M. (2011). Presenting your structures: The CCP4mg molecular-graphics software. *Acta Cryst.* D67, 386–394.

V

Where to Recruit New Enthusiasts? Science Outreach Examples

There are several different sorts of meeting the public and school-children events to explain science. I give three different examples. The first is a book that I reviewed from a researcher and museum curator in the USA. The second example is one that I took part in organised by the British Association for the Advancement of Science (BAAS). The third example was a lecture I delivered at The Royal Institution (RI) in London. The RI is another organisation besides the BAAS devoted to science outreach and with its own, famous, research laboratories.

CHAPTER 17

When and Where Are Broader Impacts of Science on Society

THE BOOK *BROADER IMPACTS OF Science on Society* by Bruce J. MacFadden, a palaeontologist, is linked to the Broader Impacts Program of the US National Science Foundation (NSF), which he became aware of whilst he was a professor and faculty curator at the Florida Museum of Natural History.[1] Clearly enthusiastic about broader impacts, he taught a graduate seminar course on the subject for seven semesters. He writes in his Preface 'I thank my students in these classes for their enthusiasm. Unlike some of my more senior colleagues, who are set in their ways, the next generation understands the need for NSF's Broader Impacts and related activities that benefit society.' In the UK, our research councils faced a similar negative reaction, even

---

[1] Adapted from my book review of *Broader Impacts of Science on Society* by Bruce J. MacFadden, Cambridge University Press, 2019 [Helliwell, J.R. (2019). *J. Appl. Cryst.* **52**, 1464–1466]. Reproduced with permission of the International Union of Crystallography.

backlash, to research impact statements in grant proposals. This is a serious culture clash seen by many scientists as an attack by the funding agencies on curiosity-driven research, naturally cherished by us scientists. Personally I was not worried and actually I was glad to see the impact of my work as a protein crystallography beamline scientist at the UK's Synchrotron Radiation Source (SRS) in providing analytical services for industry. As an enthusiastic member of our SRS annual report committee, I always regarded the benefits to industry and society narratives as golden nuggets. So, I see this as an important book from an experienced enthusiast in the parallel culture to the UK of the USA. Furthermore, the author worked for 2 years as an NSF program officer and represented NSF at the American Association for the Advancement of Science (AAAS), presenting talks on broader impacts.

'The extent to which society values basic research is at the core of the debate about return on investment of government funds.' This seemingly correct statement conceals the truth that government is a proxy for the voters, that is, taxpayers. Do governments really know in detail what taxpayers want from science and scientists? Maybe it is best in these days of 'fake news protagonists' in politics that an organisation such as the non-partisan Pew Research Center[2] does the documenting of the data on the variations of opinions of the US public and US scientists, showing big opinion gaps on topics such as GM food safety, climate change and evolution. MacFadden is incisive with his section on diversity, highlighting the poor representation of women and minorities in the USA and elsewhere, as well as the mismatch of supply of PhDs to jobs within different Science, Technology, Engineering and Medicine (STEM) sectors (except in computer sciences), and he shows explicitly how 'we will not have fulfilled our social responsibility or realised the economic benefits of diverse participation in STEM'.

---

[2] https://www.pewresearch.org/.

This book provides a clear historical description of NSF's foundation and development. He quotes from the USA NSF Foundation Act of 1950 [1], with its first sentence setting the scene for all that follows, including in effect broader impacts: 'To promote the progress of science; to advance the national health, prosperity and welfare; to secure the national defence; and for other purposes.' Annual review and oversight of NSF by both Senate and House of Representatives science committees commenced in the 1960s.

The book highlights several case studies, including of the author's work in Panama excavating 20-million-year-old fossil mammals. He wrestles with NSF policies for taking risks in encouraging adventurous research but not too much risk, thereby potentially jeopardising return on investment of the taxpayer funds they shepherd. A subtext is apparent that the riskier projects are at the small-budget end of things. The author delightfully shows a mosaic of photos showing teachers and students clearly happy to be taking part in a project about a 10-million-year-old giant shark, *Carcharocles megalodon*, and its tooth found in Panama.

In outreach matters, there is the interesting question of if and when to describe Strategic versus Curiosity Science to an audience? The author introduced me to the term 'charismatic science': if you are going to undertake a curiosity-driven topic, rather than be in a strategic (NSF) research theme, then try and select something with likely popular appeal (it will expedite one's broader impacts efforts later).

The author covers initiatives striving for Diversity, Equity and Inclusion. This is a very heterogeneous coverage of topics ranging from the situation of women in STEM (including underrepresentation, unfair pay and harassment policies), disabilities and avoiding barriers for the disabled, outreach to the prisoner population (2.2 million in the USA, including 60,000 juveniles), and underrepresented minorities in general. For the International Year of Crystallography in 2014, I approached the UK's Prisoners'

Education Trust (PET), offering a broadly based lecture on crystallography and the molecules of life, having first obtained the support of the International Union of Crystallography and the British Crystallographic Association. The PET said they would mention my offer in their newsletter, but there were no takers, and the PET explained to me that practical skills training, e.g. as chefs or car mechanics, was preferred. The USA seems to have the broader programme I had in mind as described in MacFadden's book.

The author describes teacher professional development in general, including the NSF's scheme for research experience for teachers and best practices and resources. He reflects that the overproduction of PhDs could have been turned by NSF towards helping those school districts that cannot effectively recruit and retain qualified science teachers. Specifically, 'It is unfortunate that this career pathway is not more firmly embraced by scientists and educators; it potentially would transform US workforce development education in the twenty-first century.' A very positive remark is also offered by his quotation at the start of his concluding comments of a New York teacher that, 'the single most important thing that practising scientists in any field can do is reach out to teachers and students at a local high school'.

To sum up his book, the author examined his most frequently used words in the book; for which the top five were *science, NSF, broader impacts and STEM*. Overall I found this to be an impressive book with great depth and breadth. The NSF emphasis will be of great value to those based in the USA and certainly I think of interest to the wider community of scientists, teachers and the science media at large around the world.

REFERENCE

1. Bush, V., Director of the Office of Scientific Research and Development (1945). *Science the Endless Frontier: A Report to the President.* United States Government Printing Office, Washington, DC. https://www.nsf.gov/about/history/vbush1945.htm.

CHAPTER 18

When I Lectured at the British Association for the Advancement of Science (BAAS) Festival of Science Held at the University of Liverpool in September 2008

S CIENCE FESTIVALS PRESENT AN unrivalled opportunity to educate the general public about science matters. They are particularly useful in drawing attention to scientific enterprises in the vicinity of the place where they are run. The British Science Festival of the British Association for the Advancement of Science (BAAS), Europe's longest standing science festival, travels to a different place in the UK each year. The event in Liverpool in September 2008 allowed several of us to showcase the fantastic success of the Daresbury Synchrotron in Cheshire, which had just reached the end of its life.

AFTER TWO MILLION HOURS OF SCIENCE, THE SRS BIDS FAREWELL TO THE PUBLIC[1]

A review of accelerator science was given by Dr Mike Poole, Director of the Accelerator and Technology Centre of the Science and Technology Facilities Council (STFC), of Biology and Medicine by John Helliwell, Professor of Structural Chemistry at Manchester University, of Materials Science by Bob Cernik, Professor of Materials Science at Manchester University and of Future Light Sources by Dr Tracy Turner, of the Photon Science Department of STFC (Figure 18.1).

The session was opened by Tony Buckley, Daresbury Laboratory's Head of Communications, who said,

> Cleaner fuel, safer aircraft and new medicines, not to mention a Nobel prize, great tasting chocolate and iPods - all of these things, he explained, had been influenced or made possible by the often world leading scientific research carried out on the Synchrotron Radiation Source (SRS) at the STFC Daresbury Laboratory in

---

[1] This is a shortened version of the account, but focussing on the biology and medicine theme, with specific references added, published originally in *Synchrotron Radiation News* [Cernik, B., Helliwell, J.R., Poole, M. and Turner, T. (2009). SRS highlights presented at British Association Festival of Science. *Synchrotron Radiat. News*, 22(4), 10-12, doi: 10.1080/08940880903113893] Reprinted by permission of the publisher (Taylor & Francis Ltd, http://www.tandfonline.com).

FIGURE 18.1   SRS speakers at the BA Science Festival September 2008; left to right: John Helliwell, Mike Poole, Tracy Turner and Bob Cernik.

Warrington, which formally closed on the 4th August 2008 after 28 years of operation and two million hours of science.

Focussing now on the theme of Biology and Medicine the concept for the beamlines for protein crystallography I described in Ref. [1] and brought to fruition in the X-ray beamline instruments 7.2 and 9.6 by two project teams [2,3]. The SRS's most famous achievement was the critical role it played towards a share of a Nobel Prize in chemistry to Sir John Walker in 1997 [4], for solving how the F1 ATPase enzyme works and that opened the way for new insights into metabolic and regenerative disease. The F1 ATPase enzyme is the 'molecular machine' by which protons are pumped across cell membranes, generating adenosine triphosphate (ATP), and storing energy within the cell. Also in the 1980s, Professor Michael Rossmann of Purdue University, USA, brought to the SRS crystals of human rhinovirus-14, the cause of the common cold [5]. He defined the early methods for tackling such crystal samples, which were very radiation sensitive [5], using the first SRS protein crystallography Station 7.2 [2]. Next came Harvard University, USA, with the Simian Virus 40 project, an even bigger virus studied by Professor Steve Harrison and co-workers [6] using SRS Station 9.6 [3]. Work led by Professor David Stuart and co-workers of Oxford University revealed the foot and mouth disease virus (FMDV) crystal structure [7], which even made it onto the BBC Evening News! Amongst the other highly cited work from the SRS was the light harvesting protein structures (Figure 18.2), critical to understanding the molecular basis of photosynthesis, led by Dr Miroslav Papiz of Daresbury and Professors Neil Isaacs and Richard Cogdell from Glasgow University [8,9]. The molecules causing the colouration properties of marine organisms such as lobsters were revealed in the crustacyanin crystal structures in work led by my laboratory

FIGURE 18.2 Structure determination of the light harvesting protein (based on X-ray crystallography data from the SRS [8]). This RCSB screenshot is of PDB Code 1KZU reference [9].

in collaboration with Professor Naomi Chayen, Imperial College and Dr Peter Zagalsky of Royal Holloway and Bedford New College, London, and received very large media and public interest (see also Chapter 19) [10]. Three-dimensional structural knowledge has also helped guide drug discovery; this effort was focussed via the in house Daresbury Analytical and Research Technical Services (DARTS) to the pharmaceutical industry led by Dr Pierre Rizkallah. The DARTS became the forerunner of similar analytical services to industry such as the Diamond Light Source Industrial Liaison Office (https://www.diamond.ac.uk/industry.html). Overall, some 1200 protein crystal structures were solved through experiments at the SRS and were deposited in the worldwide Protein Data Bank [11]. This very large user programme was supported through an Experiments Team led by Dr Colin Nave and latterly Dr James Nicholson; Colin also led the development of Beamline 14 [12].

Pioneering work in muscle diffraction and including leading developments of fast area detectors to study muscle contraction were undertaken at Daresbury led by Professor Joan Bordas (later Director of the ALBA synchrotron radiation source in Spain) [13]. Besides X-rays, the UV/visible photons emitted from the SRS were used to study the dichroism signals of proteins including in biological cell penetration by Dr Gareth Jones and Dr David Clarke [14]. Both such research programmes studied the whole biological system (known these days as Systems Biology) and allowed the full natural mechanism to be studied as well as medical pathologies. A third of a genome is estimated to be for genes coding for metalloproteins. Metalloproteins and the precise study by X-ray absorption spectroscopy, in complementation with protein crystallography, is another very important domain of biology and studied extensively by Professor Samar Hasnain and colleagues [15]. Another technique, powder diffraction, is used to characterise drug molecules in the tablets taken by patients. For example, the ab initio structure solution from powder data from the organic drug molecule cimetidine was done by Professor Bob Cernik and colleagues [16]. Cimetidine works by reducing the amount of acid in the stomach, used to treat occasional heartburn.

These were just a few examples, and from the biological sciences output alone, of the SRS over a remarkable 28-year period of operation which saw the facility develop from a pioneering machine to one of the most successful and useful scientific tools of the late 20th century. In terms of spin offs, the SRS also made major inputs to the technical design of the European Synchrotron Radiation Facility (ESRF) in Grenoble including the definition of the protein crystallography range of instruments led by me [17]. Daresbury also provided the ESRF Machine Advisory Committee Chair, Dr Jerry Thompson, and one of the ESRF Life Sciences Directors, Professor Peter Lindley. The Diamond Light Source in Oxford, successor to the SRS, which opened in January 2007 and at £370 million is the UK's biggest science investment in 30 years, had much technical design input on the source and the beamlines from Daresbury Laboratory.

In the worldwide race underway to design and build the next generation of SR light sources, the UK's effort in 2007 was led by the New Light Source project and ALICE, a fourth-generation prototype accelerator under development at Daresbury (https://stfc. ukri.org/research/particle-physics-and-particle-astrophysics/ alice-dl/), to generate pulses of light thousands of times shorter than can be produced from synchrotrons. This aimed at ultrafast, i.e. femtosecond studies of fast chemical and biological processes. Chapter 15 describes the currently (2020) proposed UK XFEL.

REFERENCES

1. Helliwell, J.R. (1979). Optimisation of anomalous scattering and structural studies of proteins using synchrotron radiation. *Proc. Daresbury Study Weekend* **DL/SCI/R13**, 1–6.
2. Helliwell, J.R., Greenhough, T.J., Carr, P., Rule, S.A., Moore, P.R., Thompson, A.W. and Worgan, J.S. (1982). Central data collection facility for protein crystallography, small angle diffraction and scattering at the Daresbury SRS. *J. Phys. E.* **15**, 1363–1372.

3. Helliwell, J.R., Papiz, M.Z., Glover, I.D., Habash, J., Thompson, A.W., Moore, P.R., Harris, N., Croft, D. and Pantos, E. (1986). The wiggler protein crystallography work-station at the Daresbury SRS; progress and results. *Nuclear Instrum. Methods* **A246**, 617–623.

4. Abrahams, J., Leslie, A., Lutter, R. and Walker, J.E. (1994). Structure at 2.8 Å resolution of F1-ATPase from bovine heart mitochondria. *Nature* **370**, 621–628.

5. Rossmann, M.G. and Erickson, J.W. (1983). Oscillation photography of radiation-sensitive crystals using a synchrotron source. *J. Appl. Cryst.* **16**, 629–636.

6. Liddington, R.C., Yan, Y., Moulai, J., Sahli, R., Benjamin, T.L. and Harrison, S.C. (1991). Structure of simian virus 40 at 3.8Å resolution. *Nature* **354**, 278–284.

7. Acharya, R., Fry, E., Stuart, D., Fox, G., Rowlands, D. and Brown, F. (1989). The three-dimensional structure of foot-and-mouth disease virus at 2.9 Å resolution. *Nature* **337**(6209), 709–716.

8. McDermott, G., Prince, S.M., Freer, A.A., Hawthornthwaite-Lawless, A.M., Papiz, M.Z., Cogdell, R.J. and Isaacs, N.W. (1995). Crystal structure of an integral membrane light-harvesting complex from photosynthetic bacteria. *Nature* **374**, 517–521.

9. Cogdell, R.J., Freer, A.A., Isaacs, N.W., Hawthornthwaite-Lawless, A.M., Mcdermott, G., Papiz, M.Z. and Prince, S.M. (1997). Integral membrane peripheral light harvesting complex from Rhodopseudomonas acidophila strain. *J. Mol. Biol.* **268**, 412–423.

10. Cianci, M., Rizkallah, P.J., Olczak, A., Razftery, J., Chayen, N.E., Zagalsky, P.F. and Helliwell, J.R. (2002). The molecular basis of the coloration mechanism in lobster shell: β-crustacyanin at 3.2 Å resolution. *Proc. Nat. Acad Sci. USA* **99**(15), 9795–9800.

11. Helliwell, J.R. (2012). The evolution of synchrotron radiation and the growth of its importance in crystallography. *Crystallogr. Rev.* **18**(1), 33–93.

12. Duke, E.M., Kehoe, R.C., Rizkallah, P.J., Clarke, J.A. and Nave, C. (1998). Beamline 14: A new multipole wiggler beamline for protein crystallography on the SRS. *J. Synchrotron Radiat.* **5**(Pt 3), 497–499.

13. Bordas, J., Diakun, G.P., Diaz, F.G., Harries, J.E., Lewis, R.A., Lowy, J., Mant, G.R., Martin-Fernandez, M.L. and Towns-Andrews, E. (1993). Two-dimensional time-resolved X-ray diffraction studies of live isometrically contracting frog sartorius muscle. *J. Musc. Res. Cell Mot.* **14**, 311–324.

14. Clarke, D.T. and Jones, G. (2004). CD12: A new high-flux beamline for ultraviolet and vacuum-ultraviolet circular dichroism on the SRS, Daresbury. *J. Synchrotron Radiat.* **11**(Pt 2), 142–149.

15. Ascone, I., Fourme, R., Hasnain, S.S. and Hodgson, K.O. (2005). Metallogenomics and biological X-ray absorption spectroscopy. *J. Synchrotron Radiat.* **12**, 1–3.

16. Cernik, R.J., Cheetham, A.K., Prout, C.K., Watkin, D.J., Wilkinson, A.P. and Willis, B.T. M. (1991). The structure of cimetidine ($C_{10}H_{16}N_6S$) solved from synchrotron-radiation X-ray powder diffraction data. *J. Appl. Cryst.* **24**, 222–226.

17. Helliwell, J.R. (1987). The ESRF Foundation Phase Report, European Synchrotron Radiation Facility (ESRF), Grenoble, 615 p.

CHAPTER 19

When I Presented a Discourse at the Royal Institution

T HE FRIDAY EVENING DISCOURSES are a prestigious event for any scientist to be invited to give. They are described by The Royal Institution (RI) in the following way:

> Discourses are one of the RI's oldest and most prestigious series of talks. Since 1825, audiences in the theatre have witnessed countless mind-expanding moments, including the first public liquefaction of air by James Dewar, the announcement of the electron by J. J. Thomson and over 100 lectures by Michael Faraday. In more recent times, we have had Nobel laureates, Fields medal winners, scientists, authors and artists – all from the cutting-edge of their field. Discourses are an opportunity for the best and brightest to share their work with the world.
>
> Steeped in nearly two centuries of tradition, a Discourse is more than just a lecture. To keep the focus

on the topic, presenters begin sharply at 7:30 pm without introduction and we lock the speaker into a room ten minutes ahead of the start (legend has it that a speaker once tried to escape!). Some of our guests dress smartly for our Discourse events to add to this sense of occasion.

The Friday Evening Discourse that I presented in 2003 (Figure 19.1a and b) was entitled *Why does a lobster change colour on cooking?* It is absolutely true that I was locked in a room ten minutes ahead of the start of my lecture.

I was glad to have Dr Peter Zagalsky attend my lecture (see Figure 19.1b). Peter's work on the purification of the proteins responsible for the colouration of the marine invertebrates *Homarus gammarus* (European lobster), *Homarus americanus* (North American lobster) and *Velella velella* (common names: the sea-raft or by-the-wind sailor) commenced in the 1960s. In his very first publication [1], entitled *Purification and Properties of Crustacyanin* – an apt title of his life's work – he was able to isolate from a carapace (Figure 19.2) extract a blue fraction which showed complete homogeneity in electrophoresis on starch gel. Further work in the early 1960s with collaborators [2-5] used spectroscopic, biophysical and chemical methods to obtain information on their structure. This area remained the major focus of his lifetime's research.

Peter went on to work with a large number of scientists besides myself. The amino acid sequence of crustacyanin was determined in a collaboration with Professor John B.C. Findlay's team, University of Leeds [6]. The crystallisation of the β-crustacyanin protein proved to be immensely difficult but was ultimately achieved by Professor Naomi Chayen at Imperial College, London. This enabled the determination of its structure by X-ray crystallography by my team at the University of Manchester and the Synchrotron Radiation Source at the Daresbury Laboratory, Warrington [7], based on the apocrustacyanin A1 structure solved with softer X-rays from the Daresbury synchrotron [8].

FIGURE 19.1   At the Royal Institution, London, to present my Friday Evening Discourse outreach lecture on *Why do Lobsters Change Colour on Cooking?* (a) With my wife and sons James and Nick and collaborators and friends in the RI Library and (b) with the Director of the Royal Institution Baroness Susan Greenfield and my collaborator, on the right, Dr Peter Zagalsky.

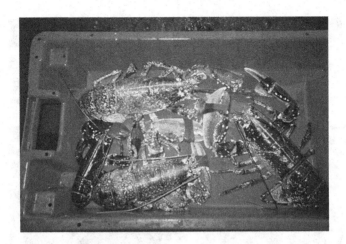

FIGURE 19.2   Lobsters caught by fisherman at the Isle of Whithorn, Scotland. (My own photograph.)

The resulting X-ray crystal structure of β-crustacyanin provided an explanation for why a lobster changed colour when boiled. The carotenoid prosthetic group of the protein, astaxanthin, is orange-red but assumes a blue-black colour when bound to crustacyanin. On boiling, it can be readily imagined then that the denaturation of the holoprotein results in the release of the astaxanthin, accounting for the colour change, known as the bathochromic shift [7]. The publication [7] attracted a huge amount of interest not only by scientists but also by the media – newspapers, TV, radio stations worldwide, including the BBC's Today radio programme and a thankyou letter from the journal where we published the work, the Proceedings of the National Academy of Sciences. The journal *Nature*, where we had submitted to first, had rejected our article because of insufficient general interest and that a specialist journal would be more appropriate!

As an aside, a thorny issue and debate at its peak during the years of my graduate studies (1974–1977) was the relevance of the protein crystalline state to the solution state of a protein inside the biological cell. Nuclear magnetic resonance (NMR) studies provided atomically detailed results in solution and of course protein crystallography provided atomic details for a protein in the solid state. The crustacyanin studies that I described in my Discourse were very relevant to that debate decades earlier. The α- or β-crustacyanin crystals, when frozen, are still blue as they are blue at room temperature. There is also a concern these days about cryo-artefacts, as so many protein crystal structures are done at 100 K, to slow down X-ray damage to the crystal sample. But as more crystal structures have been done at both cryo and at room temperature, there are types of differences in the protein structures revealed. In the case of the β-crustacyanin crystal structure, we also measured the small-angle X-ray scattering (SAXS) on β-crustacyanin solution at room temperature and the fit of the β-crustacyanin cryo crystal structure [8] was a perfect fit to the SAXS curve [9]. So, the room temperature and cryo β-crustacyanin structures are the same.

The Royal Institution Friday Evening Discourse is a national tradition, a big honour I felt. Afterwards my wife and I were hosted to dinner by Baroness Greenfield and we were joined by the radio programme host, Edward Stourton. I could then also reasonably hope that the science I presented would lead to further 'science in the media'.

REFERENCES

1. Zagalsky, P.F. (1963). Purification and properties of crustacyanin. *Biochem. J.* **89**, 21P2.
2. Cheesman, D.F., Zagalsky, P.F. and Ceccaldi, H.J. (1966). Purification and properties of crustacyanin. *Proc. R. Soc. London, Ser. B* **164**, 130–151.
3. Cheesman, D.F., Lee, W.L. and Zagalsky, P.F. (1967). Carotenoproteins in invertebrates. *Biol. Rev.* **42**, 131–160.
4. Gammack, D.B., Raper, J.H., Zagalsky, P.F. and Quarmby, R. (1971). The physical properties of the lobster carapace carotenoprotein crustacyanin. *Comp. Biochem. Physiol. Part B Comp. Biochem.* **40**, 295–300.
5. Quarmby, R., Norden, D.A., Zagalsky, P.F., Ceccaldi, H.J. and Daumas, R. (1977). Studies on the quaternary structure of the lobster exoskeleton carotenoprotein, crustacyanin. *Comp. Biochem. Physiol. Part B Comp. Biol.* **56**, 55–61.

6. Keen, J.N., Caceres, I., Eliopoulos, E.E., Zagalsky, P.F. and Findlay, J.B. (1991). Complete sequence and model for the C1 subunit of the carotenoprotein, crustacyanin, and model for the dimer, β-crustacyanin, formed from the C_1 and A_2 subunits with astaxanthin. *Eur. J. Biochem.* **202**, 31–40.

7. Cianci, M., Rizkallah, P.J., Olczak, A., Chayen, N.E., Zagalsky, P.F. and Helliwell, J.R. (2002). The molecular basis of the coloration mechanism in lobster shell: β-crustacyanin at 3.2 Å resolution. *Proc. Natl. Acad. Sci. USA* **99**, 9795–9800.

8. Cianci, M., Rizkallah, P.J., Olczak, A., Raftery, J., Chayen, N.E., Zagalsky, P.F. and Helliwell, J.R. (2001). The crystal structure of lobster Apocrustacyanin A1 using softer X-rays. *Acta Crystallogr.* **D57**, 1219–1229.

9. Chayen, N.E., Cianci, M., Grossman, G., Habash, J., Helliwell, J.R., Nneji, G.A., Raftery, J., Rizakallah, P. and Zagalsky, P.F. (2003). Unravelling the structural chemistry of the colouration mechanism in lobster shell. *Acta Crystallogr.* D **59**, 2072–2082.

VI

Big Issues for Scientists and for Members of the Public and Schoolchildren

CHAPTER 20

Where to Place One's Trust?

A S SOMEONE WHO HAS THRIVED in a scientific life, of course I would say 'there is trust in science'.[1] But in these increasingly sceptical times, scientists are having to place more energy into convincing the public, and some politicians, of this. My recent review of the book *Why Trust Science?* provided the opportunity for me to describe some thoughts on this topic.

Naomi Oreskes is a very accomplished 'American historian of science', also previously a geochemist in the mining industry.[2] The title of her book *Why Trust Science?* poses an all important question for everyone. As I was reviewing this book, the National Academies of Science, Engineering and Medicine (2019) in the USA published their 2019 report [1] on *Reproducibility and*

---

[1] Based on a review of the book *Why Trust Science?* By Naomi Oreskes. Princeton University Press, 2019, [Helliwell, J. R. (2019). *J. Appl. Cryst.* **52**, 1461–1463]. Reproduced with permission of the International Union of Crystallography.

[2] See her biography at https://en.m.wikipedia.org/wiki/Naomi_Oreskes. There is also a taster for the book with her excellent 2014 TED Talk (https://tedsummaries. com/2014/08/03/naomi-oreskes-why-we-should-trust-scientists/).

Replicability in Science, a 256-page analysis and survey including a final chapter of 20 pages on 'Confidence in Science'. Whilst naturally USA focussed, it also had international participation. I found this an authoritative report with good recommendations, including for journals and societies. Data and software availability featured prominently in the best processes for achieving as high a scientific standard as possible. The cogency of this report has greatly informed my review.

Why Trust Science? is an excellent format for a book; as well as the core argument written by Naomi Oreskes herself, it includes several chapters offering written critiques, with Naomi Oreskes' responses. The book opens with reviews of some recent examples of disputed science. The first is the issue of some public disquiet, fanned by President Trump, about the vaccination of schoolchildren, which she sums up by finally stating the data of the situation: the number of cases of the diseases in question has gone up as a result of vaccination avoidance. Example 2 again involves the Trump administration, this time the Vice President's belief in creationism, backed up by a poll of US churchgoers where 67% would espouse the same view. The third and final example for her Chapter 1 is again USA based and concerns the American Enterprise Institute, which apparently had offered cash incentives to find errors in the Intergovermental Panel on Climate Change's (IPCC's) analyses. The author sets her overall metric for monitoring science as a process: there has to be a consensus. She shows no sense of immediate danger in that metric. For example, the consensus among the physics community at the end of the 19th century was generally that there was nothing major left to discover, it only required details to be filled in. How wrong that consensus was! It needed piercing intellects such as of Albert Einstein, Max Planck and so on to radically overhaul that false consensus. The tests of their theories of relativity and a quantum world rested on experiments and in turn provided the new scientific data of the time. I will return to the topic of consensus below.

Her section entitled Getting Unstuck: Social Epistemology introduces the criticism of science that it claims to be objective and yet it largely excludes half the population, females! Yes, how very true; science needs to become more representative. My efforts to do something about gender disparity I describe in my book *Skills for a Scientific Life* [2].

The author raises an interesting and indeed fundamental point. Paraphrasing, if science progresses *via* a process of falsifiability *à la* Karl Popper [3], then why should the public trust a science result now? In my own book *The Whats of a Scientific Life* [4], I answer that science delivers, whether it is health, wealth or comfort, modes of travel, smartphones or the internet. Such examples I think provide answers to the author's question of how we can know which science truths will be permanent. I see no need for a distinction between pure and applied science, nor did Max Perutz in his marvellous book *Is Science Necessary?* [5].

In the chapter entitled 'Pascal's Wager Reframed: Towards Trustworthy Climate Policy Assessments for Risk Societies', by Professor Ottmar Edenhofer (an expert on climate change policy) and Dr Martin Kowarsch (an expert on environmental policy), the topic of the public's reaction to climate change is addressed or rather its diverse reactions to it. Pascal's wager is that 'it is in one's own best interest to behave as if God does exist, since the possibility of eternal punishment in hell outweighs any advantages in believing otherwise'. This chapter focusses then on climate science and the consequent step of determining a policy of what to do about it. This is a very useful distinction. As I have quoted Winston Churchill before about scientists and who said 'they should be on tap but not on top'. I challenged that statement of Churchill in my book *Skills for a Scientific Life* [6] by saying that a policy committee needs to include an expert scientist, not only to advise but as a voting member. Edenhofer and Kowarsch describe the spectrum of consequences of climate change, which range from 'the Trump administration's estimate of the social cost of carbon of $1 to $6 per additional ton of carbon dioxide into the atmosphere to $45 as the estimate by the Obama Presidential administration', to which I add the Extinction Rebellion estimate of consequences if we do not achieve net zero emissions by 2025 as '6 billion deaths due to mass starvation, that is what the science is telling us' [7]. The flaw in that BBC programme format [7], and many others of its type, is that they have an interviewer, Stephen Sackur, and an advocate for the science, Roger Hallam, but not a lead scientist such as the Chair of the IPCC. Overall I found that this chapter really only set the scene, as it did not summarise the IPCC report, so as to dissect the contrasting policy positions of the Trump and Obama presidential administrations, which are quoted, let alone consider the extrapolation made by Extinction Rebellion that I cite. This chapter could have been pivotal but must await another edition of Naomi Oreskes' book.

Inspired by Naomi Oreskes, Professor Jon A. Krosnick, a practitioner of science in social psychology and cofounder of Best Practices in Science at Stanford University comments on the present and future of science. He concluded with suggestions for remedies to avoid irreproducibilities in future. Inevitably the National Academies report recommendations are deeper, wider and authoritative. Furthermore, the latter emphasise the importance of archiving and transparency of data, as well as of software, basically following the FAIR (Findable, Accessible, Interoperable and Reusable) and FACT (Fair, Accurate, Confidential (where necessary) and Transparent) data principles [8,9].

Naomi Oreskes replies to the other authors. She agrees with the separation of science from technology, which I disagreed with as a position already above. She agrees with the need for consensus, which I have argued above is not the main point because it is the underpinning data that make or break a science publication. She disagrees with climate science and policy being connected on the basis that it is not only the science that determines policy. Hers is a controversial position, I think, that she adopts with that view. My answer is to have committees and panels determining policy to include scientists as voting members. I see that I need to make my assertion stronger still, given Naomi Oreskes' position on this; scientists need to not just be voting members and expert advisors but have a veto on policy plans if the science demands it.

On the views of the replication crisis in science by Jon A. Krosnick, Naomi Oreskes incisively states that 'Professor Krosnick makes broad claims on limited evidence and lumps together phenomena that may be distinct.' Furthermore, she correctly states that 'What leads to reliable scientific knowledge is the process by which claims are vetted.' To which I would add the process needs to include pre- and post-publication peer review and involve scrutiny of the words of the article with the data and of course requires the archiving of those data for post-publication review. So, I dispute her notion that 'a single paper cannot be the basis for reliable scientific knowledge'. She does, however, mention the need for open reporting of data.

Overall, I found this book to be a marvellous, up to date, thorough historical survey of science and its processes. The odd thing about the book is its choice of some of the modern science topics. Naomi Oreskes describes well the improvement that we are seeing of science as a process. But we are also seeing the problems of predatory journals and conferences, a few of which can seem plausible enough, and we are also realising the publications bias that is the reporting only of positive, not null or repeat, results. Her final remark is: 'confidence in science is warranted whereas the scepticism in scientists' findings in their domain of expertise is unwarranted'. I would add that even in one's own scientific domain, findings must be accompanied by their underpinning data, a theme I explore further in the next Chapter 21.

REFERENCES

1. National Academies of Sciences, Engineering, and Medicine (2019). *Reproducibility and Replicability in Science*, The National Academies Press, Washington, DC. doi: 10.17226/25303.
2. Helliwell, J.R. (2017). Chapter 27 in *Skills for a Scientific Life*, CRC Press, Boca Raton, FL.

3. Karl Popper *Logik der Forschung* (1934, revised and translated into English in 1959 as *The Logic of Scientific Discovery*) published by Routledge, Abingdon, UK.
4. Helliwell, J.R. (2020). Chapter 1 in *The Whats of a Scientific Life*. CRC Press/Taylor & Francis, Boca Raton, FL.
5. Perutz, M. (1991). *Is Science Necessary? Essays on Science and Scientists*, Oxford University Press, Oxford, UK.
6. Helliwell, J.R. (2017). Chapter 31 in *Skills for a Scientific Life*, CRC Press, Boca Raton, FL.
7. Hallam, R. (18th August 2019) on the BBC's Hard Talk, https://www.bbc.co.uk/programmes/w3csy93l.
8. Wilkinson, M.D., Dumontier, M., Aalbersberg, I.J., Appleton, G., Axton, M., Baak, A., Blomberg, N., Boiten, J.-W., da Silva Santos, L.B., Bourne, P.E., Bouwman, J., Brookes, A.J., Clark, T., Crosas, M., Dillo, I., Dumon, O., Edmunds, S., Evelo, C.T., Finkers, R., Gonzalez-Beltran, A., Gray, A., Groth, P., Goble, C., Grethe, J.S., Heringa, J., Hoen, P.A.C., Hooft, R., Kuhn, T., Kok, R., Kok, J., Lusher, S.J., Martone, M.E., Mons, A., Packer, A.L., Persson, B., Rocca-Serra, P., Roos, M., van Schaik, R., Sansone, S.-A., Schultes, E., Sengstag, T., Slater, T., Strawn, G., Swertz, M.A., Thompson, M., van der Lei, J., van Mulligen, E., Velterop, J., Waagmeester, A., Wittenburg, P., Wolstencroft, K., Zhao, J. and Mons, B. (2016). The FAIR guiding principles for scientific data management and stewardship. *Sci. Data*, **3**, 160018.
9. van der Aalst, W.M.P., Bichler, M. and Heinzl, A. (2017). Responsible data science. *Bus. Inf. Syst. Eng.* **59**, 311–313.

CHAPTER 21

Where Are the Primary Experimental Data? Ensuring Reproducibility in Science

A MAJOR TREND IN COMPUTING is that the capacity of memory store and disk space store these days has greatly, greatly increased. Also we live quite happily with terms like Big Data. What does this huge change in data storage capacity mean for the process of science? In crystallography, I was asked to lead a Working Group of the International Union of Crystallography (IUCr)[1] to

---

[1] The Diffraction Data Deposition Working Group, DDDWG (https://www.iucr.org/resources/data/dddwg).

consider what the new data archive possibilities meant for our field of research. We had excellent archives for the fruits of our research, the atomic coordinates in our molecules. These started more than 50 years ago by Dr Olga Kennard OBE FRS (who I also highlighted in my introduction to Part 2) when data storage opportunities were much, much less. The concept of linking our data to our research articles has been embedded in our crystallography community from our early days. As data archiving expanded, we as a community were able to also preserve our processed diffraction data. But processing meant subjectively making decisions of what measured data to include in our calculations. Science is known as objective. But is it objective if we don't preserve our primary experimental data, the raw data? Data archives now offer sufficient capacity that for many fields of science, including crystallography, raw data that we measure can also be preserved. Our DDDWG Report took 6 years to consult the global community, then prepare and deliver it to the Congress and General Assembly of the IUCr held in Montreal in 2017. We had been convened in 2011 at the earlier Congress and General Assembly of the IUCr held in Madrid. The final report can be found here [1]. We concluded that such raw data archiving was feasible and should be strongly encouraged. The Recommendations were to be conveyed via the IUCr Executive Committee who endorsed them to the large number of IUCr Commissions for them to consider implementation in their respective areas.

The top two Recommendations were quite generic and which were:

- Authors should provide a permanent and prominent link from their article to the raw data sets which underpin their journal publication and associated database deposition of processed diffraction data (e.g. structure factor amplitudes and intensities) and coordinates, and which should obey the 'FAIR' principles, that their raw diffraction data sets should be Findable, Accessible, Interoperable and Re-usable (https://www.force11.org/group/fairgroup/fairprinciples).

- A registered Digital Object Identifier (doi) should be the persistent identifier of choice (rather than a Uniform Resource Locator, url) as the most sustainable way to identify and locate a raw diffraction data set.

Articles describing best practice were also published [2,3].

The Commission on Biological Macromolecules has been the earliest to publish an implementation, which they did in conjunction with myself as Chair of the IUCr Committee on Data [4]. The IUCr Commission on Structural Chemistry has organised a discussion workshop to be held in Prague in 2021[2] to take community feedback on the DDDWG Report and Recommendations.

Overall, the question "Where are the primary experimental data?" is a hugely important one and their archiving is the means for ensuring objectivity, or the closest we can come to it, and also ensuring reproducibility in science. What do I mean by "the closest we can come to it"? Our measuring apparatus has to be calibrated, and this does require a trained person to ensure that the apparatus is indeed functioning properly. That person calibrating the apparatus means there is still a trace of subjectivity left in the scientific process. But with the raw diffraction data available in any study, it offers readers of a piece of research the chance to repeat any of the steps in data processing and interpretation that the original researchers undertook. That can be with other software or by altering the set parameters in the researchers' preferred software. Any variations in results are then in the hands of the future users of a piece of research. As The Royal Society's motto neatly puts it; *Nullius in verba* (take nobody else's word for it).

---

[2] https://www.iucr.org/resources/data/commdat/prague-workshop-cx.

REFERENCES

1. Helliwell, J.R., McMahon, B., Androulakis, S., Szebenyi, M., Kroon-Batenburg, L.M.J., Terwilliger, T.C., Westbrook, J. and Weckert, E. (2017). Final report of the IUCr Diffraction Data Deposition Working Group. https://www.iucr.org/resources/data/dddwg/final-report.
2. Kroon-Batenburg, L.M.J., Helliwell, J.R., McMahon, B. and Terwilliger, T.C. (2017). Raw diffraction data preservation and reuse: Overview, update on practicalities and metadata requirements. *IUCrJ* **4**, 87–99.
3. Helliwell, J.R., McMahon, B., Guss, J.M. and Kroon-Batenburg, L.M.J. (2017). The science is in the data. *IUCrJ* **4**, 714–722.
4. Helliwell, J.R., Minor, W., Weiss, M.S., Garman, E.F., Read, R.J., Newman, J., van Raaij, M.J., Hajdu, J. and Baker, E.N. (2019). Findable Accessible Interoperable Re-usable (FAIR) diffraction data are coming to protein crystallography. *Acta Cryst.* **D75**, 455–457.

CHAPTER 22

When to Cross the Boundaries of Different Academic Disciplines

I HAVE BEEN FORTUNATE IN MY career to work in a very interdisciplinary way.[1] Indeed, in elucidating the nature of the material world, science cannot be bound by man-made classifications and practices. Yet the contemporary culture of science does not always understand this simple fact. A recent book by Catherine Lyall, *Being an Interdisciplinary Academic: How Institutions Shape University Careers*, provides much food for thought on how current attitudes regarding disciplinary silos can affect the scientific life.

---

[1] Adapted from my book review of *Being an Interdisciplinary Academic: How Institutions Shape University Careers*, by Catherine Lyall. Palgrave Pivot, 2019. Hardback pp. 154. Price EUR 51.99, ISBN 978-3-030-18658-6 [Helliwell, J R. (2020). *J. Appl. Cryst.* **53**, 596–597). Reproduced with permission of the International Union of Crystallography.

I wanted to review this book as my scientific career has been across the disciplines, be it my crystallography or my biophysics research, and where possible its applications, especially of instrumentation and methods development for the best chemistry analytical science.

Catherine Lyall, originally a graduate chemist, is a very experienced researcher into the topic of interdisciplinary research and someone who has provided reports on the topic for various important funding and university organisations. The book has a sub-text I think which is one of eager youngsters trying to make their careers versus the bias of heads of single subject departments. Meanwhile society looks on at the neglect of its pressing problems whilst it perceives that the focus is on fine details within academic disciplines. So this book is very much a book for our time, and I highly recommend it.

The challenge tackled by Catherine Lyall is underpinned firstly by her interviews of those who practise interdisciplinary research and her considerable personal experience. She pointedly states '(there) is a manifest misalignment (between interdisciplinary and single discipline research) … The prevailing norms being discipline-based scholarship'. But why can some scientists be so against interdisciplinary research? There is a fear of an erosion of the research funds for their discipline-driven university departments. I tried to counter this fear in a posting at *Nature* that interdisciplinary science can pull in new research funds to science and its cash-strapped scientific researchers [1]. Whilst these points document the big issues, the overall direction of travel is obvious with the merger in 2018 of the International Council of Science (formerly the International Council of Scientific Unions) and the International Social Sciences Council to form the International Science Council.

Lyall's interviews, 22 of them, reveal the career challenges that have been faced by interdisciplinary-driven researchers and showed up the need 'to improve on current practice…especially of governance'. Then there were in addition ten leaders of universities who were interviewed. Lyall confides that a weakness of the interviewees as a group is their being UK or northern Europe based. That said, I found that her narrative and approach is well balanced by the references lists, and for example for her Chapter 1 includes policy documents from the Global Research Council, the US National Science Foundation and on Australian research governance.

The researchers she interviewed were a cohort of people originally trained some 10-15 years earlier during their PhDs on two schemes devoted to interdisciplinary research. One funding scheme was the UK's Economic and Social Research Council (ESRC) with the Natural Environment Research Council (NERC) and the other was the ESRC with the Medical Research Council (MRC). The interviewees describe their 'identity fatigue' whose solution was focussing on their research areas rather than an academic subject. The references list cites two extensive evaluation reports authored by Lyall and coauthors [2,3].

Lyall sharply scrutinises departmental cultures. Shocking anecdotes are described such as the interviewee who had been channeled by the funding body to face one of its funding panels (biology) and was asked 'Are you one of us?' Also, she describes the researcher whose interdisciplinary publication was rated by their university as poor, and impeded their promotion, but for their coauthor from a different subject their university had rated the same publication highly. Such research assessment inconsistencies confirm the extra hazards of being judged by single discipline-based peers. The question is 'What could be done to mitigate the negative consequences? on an interdisciplinary researcher's career'. The references list includes further useful reports [4,5].

In particular, she addresses the question of 'when should one become involved in interdisciplinary research?' She canvasses views of her selected PhD cohorts and university leaders as well as funding agencies. Her interdisciplinary research PhDs remained enthusiastic about their training with a few, modest, provisos whereas university leaders stressed the need for a person to be strong in a discipline before joining interdisciplinary research projects. Either way it is a person who 'can see interconnections between disciplines and bring them together synergistically'. She concludes that such a team leader for broad-based research challenges may fare better in their career development in industry or government science centres than the largely vertically arranged discipline alongside discipline in a university.

In trying to understand the preferences of individual scientists for their work choices, she quotes Isaiah Berlin (based on Archilochus from 700 BC) [6] of scholars being either single discipline 'hedgehogs' or interdisciplinary 'foxes'. In terms of what steps institutions themselves should take to develop interdisciplinarity in its staff, a difficulty is that university leaders themselves do not have a consensus. Lyall summarises that these leaders show 'fundamental misunderstandings about the nature of interdisciplinary knowledge, how this is acquired, and the skills that interdisciplinary researchers offer'. Furthermore, society at large should indeed be troubled about the universities claiming to be for the public good and yet showing such 'fundamental misunderstandings'. A useful overview report is from the League of European Research Universities (LERU 2016) [7] involving a panel of 23 research intensive universities.

I do offer a criticism. So as to inspire the reader better, the book could have focussed more on specific research challenges. As I explain in Chapter 16 of my recent book *The Whats of a Scientific Life*, major challenges such as climate change [8], or ageing, do not respect academic boundaries. These can and do inspire activities such as for the UK Global Challenges Research Fund[2] and are mentioned by Lyall. However, they can and should also be set in the context of single-discipline research examples chosen to inspire, such as the big bang in astronomy or chemical catalysis, which are wonderful examples of single-discipline research.

---

[2] See https://www.ukri.org/research/global-challenges-research-fund/.

In summary, I liked this book, not least because of its data gathering from interviews of two decent-sized cohorts of interviewees, namely the practitioners and the senior university managers. Also it is set in the context of the formal work done by the extensive personal work of the author for various research bodies. The book also finishes well with proposals for both practical and institutional governance reform. I found the book well written and with extensive references for each chapter, various of which I accessed via the web as pdfs, including the official reports. If Lyall's well-researched recommendations are followed, her book will have a very positive impact.

REFERENCES

1. Helliwell, J.R. (2007). Interdisciplinary research could pull cash into science. *Nature* **448**, 533. doi: 10.1038/448533b.
2. Meagher, L. and Lyall, C. (2005). Evaluation of the ESRC/NERC interdisciplinary research studentship scheme. https://esrc.ukri.org/files/research/research-and-impact-evaluation/esrc-nerc-interdisciplinary-research-studentship-scheme/.
3. Meagher, L. and Lyall, C. (2009). Evaluation of the ESRC/MRC interdisciplinary studentship and postdoctoral fellowship scheme. https://esrc.ukri.org/files/research/research-and-impact-evaluation/esrc-mrc-interdisciplinary-studentship-and-postdoctoral-fellowship-scheme/.
4. Lyall, C. and King, E. (2013). *International Good Practice in the Peer Review of Interdisciplinary Research Report of a Scoping Study Conducted for the RCUK Research Group*, The University of Edinburgh. https://www.research.ed.ac.uk/portal/files/23461807/Lyall_and_King_Interdisciplinary_Peer_Review.pdf.
5. National Academy of Sciences, National Academy of Engineering, and Institute of Medicine (2005). *Facilitating Interdisciplinary Research*, The National Academies Press, Washington, DC. doi: 10.17226/11153.
6. Berlin, I. (1953). *The Hedgehog and the Fox: An Essay on Tolstoy's View of History*, Weidenfield and Nicholson, London, UK.
7. LERU (League of Research Intensive Universities) (2016). *Interdisciplinarity and the 21st Century Research-Intensive University*, LERU, Leuven, Belgium.
8. Helliwell, J.R. (2020). *The Whats of a Scientific Life*. CRC Press Taylor & Francis, Boca Raton, FL.

VII

The Role of Scientists Themselves, the Funding Agencies and Government in Determining the Future in Research and Development: a Very Important When and Where

CHAPTER 23

Where Are the Scientists Heading?

SOMETIMES A SCIENTIST PROJECTS into the future. This I volunteered to do at the International Union of Pure and Applied Biophysics (IUPAB) Congress and General Assembly in 1999 held in New Delhi, India, in September 1999. The final session of the Congress was entitled 'Biophysics in the 21st Century' in which I and two other scientists presented a lecture. Figure 23.1 shows these three abstracts, kindly provided to me by Professor Tony North of the University of Leeds who, unlike me, had preserved his IUPAB 1999 Abstracts book. Notice that while I ranged over a number of specific developments, the bioinformaticists would only hazard predictions into the following year, while the biophysicist took a much more expansive view.

My lecture was quite focussed on large-scale centralised technology infrastructures; the synchrotron X-ray sources, the neutron sources and the X-ray lasers. I reproduce my lecture abstract (Figure 23.1) as text:

> Next century synchrotron X-radiation and neutron protein crystallography in molecular biophysics

S54

Biophysics will prosper in the 21st century

Charles R. Cantor, CSO, Sequenom Inc San Diego CA USA and Sequenom GmbH, Hamburg Germany

It was Walter Gilbert who first pointed out that biology was about to go a paradigm shift. He argued persuasively that the massive increase in available DNA sequence data would result in a decreased reliance on biological experimentation and an increased use of database searches and computer simulations that would either replace experiments or act as a prelude to them.

As the 21st century dawns, we will see complete genome sequences for humans and many other species. We are also likely to see a complete enough survey of protein and nucleic acid structural motifs that computation will replace experiments in most macromolecular structure determination. The continuing increase in computer power available almost for free to all biologists will drive a revolutionary change in the way scientific research and all other aspects of science are carried out. For example, mapping of DNA sequence polymorphisms in many species is already easier in silico than experimentally. High throughput analytical tools like automated mass spectrometry or expression profiling using DNA chips require computation both for experimental design and for analysis of the resulting deluge of data.

However the need for experiments will not end. In fact in one of the areas likely to be key for 21st century biology, there is currently an enormous experimental backlog. It is clearly a desirable goal to be able to make accurate simulations of the behavior of single cells. Eventually these need to be scaled to predict the behavior of assemblies of cells, even whole organisms. To carry out such simulations enormous numbers of in vitro reaction rate constants will have to be measured. It is not at all clear that rates measured in vitro will be adequate. We need to develop the technology to permit such measurements.

The average biologist today is almost totally unfamiliar with the tools of modern mathematics. These tools underlie much of computationally based biology. Without knowledge of these tools it is unlikely that innovative advances in the use of computers in biology will be possible. Hence the biophysicist (and the biomedical engineer) seem poised to play a much more central role in major biological advances as these become increasingly more dominated by computation.

S55

Bioinformatics in the 21st Century
Barry Robson+ and Jean Garnier*
*Unité de Bioinformatique, Biotechnologies, INRA, Jouy-en-Josas, France
+J.T. Watson Research Center, IBM, NY, USA

Through 2001 will see a breakdown of the artificial boundaries between the disciplines. Many with biophysics training will turn to 'three dimensional molecular genetics'(3D-MG). The area will involve studying protein variation in human individuals and populations, variation of proteins in pathogens, and will provide insight for protein engineering and development of new engineering processes and new variants of organisms. In Health Care, there will be applications for both "fast response molecular medicine" and "personalized molecular medicine". As input to such calculations, one can expect a daily avalanche of data of several hundred sequences a day. Therefore, one foresees researchers performing currently multiteraflop computation and handling petabytes of data through fast transmitting devices from multiple resources, with 3D-MG involving high degrees of automation of protein modeling.

S56

Next century synchrotron X-radiation and neutron protein crystallography in molecular biophysics

John Helliwell, Department of Chemistry, University of Manchester, M13 9PL, UK

The capabilities for crystal structure studies in molecular biophysics, are being transformed by better technique and improving technology. Synchrotron X–radiation (SXR) capability is beginning to allow genome level of numbers of protein structure determinations and has opened up time–resolved Laue diffraction structure studies. Studies on the enzyme hydroxymethylbilane synthase, using MAD and Laue diffraction, will be described as examples. Ultra–high resolution SXR protein crystallography and neutron Laue diffraction is better defining hydrogen/deuterium positions and dynamics. As illustration the structures of the lectin protein concanavalin A (MW 25 kDa) studied at 0.9 Å resolution by SR X–ray and of D2O–concanavalin A by neutron Laue diffraction at 2.4 Å resolution, and the X–ray structures of sugar complexes with concanavalin A will be compared. Molecular dynamics trajectory calculations for this protein show how the structure and thermodynamics of ligand binding can be linked. Proposed X–ray SR sources and pulsed neutron sources have machine life expectancies to the middle of the next century. Extrapolation beyond includes a fully dedicated 'protein crystal genomics synchrotron' and the 'XFEL' for sub–picosecond time–resolved biophysics experiments.

30

FIGURE 23.1 Three abstracts in the biophysics in the next century concluding session of the IUPAB 1999.

The capabilities for crystal structure studies in molecular biophysics, are being transformed by better technique and improving technology. Synchrotron X-radiation (SXR) capability is beginning to allow genome level of numbers of protein structure determinations and has opened up time-resolved Laue diffraction structure studies. Studies on the enzyme hydroxymethylbilane synthase, using Multiple Wavelength Anomalous Dispersion (MAD) and Laue diffraction, will be described as examples. Ultra-high resolution Synchrotron X-ray Radiation (SXR) protein crystallography and neutron Laue diffraction are better defining hydrogen/deuterium positions- and dynamics. As illustration the structures of the lectin protein concanavalin A (Molecular Weight 25 kDa) studied at 0.9 Å resolution by SR X-ray and of D_2O-- concanavalin A by neutron Laue diffraction at 2.4 Å resolution, and the X-ray structures of sugar complexes with concanavalin A will be compared. Molecular dynamics trajectory calculations for this protein show how the structure and thermodynamics of ligand binding can be linked. Proposed X-ray SR sources and pulsed neutron sources have machine life expectancies to the middle of the next century. Extrapolation beyond includes a fully dedicated 'protein crystal genomics synchrotron' and the X-ray Free Electron Laser (XFEL) for sub-picosecond time-resolved biophysics experiments.

So, how could I be so certain of my projections? In the topics that I chose, these facilities take quite a long time to build, say 3–5 years. They then operate their science user research programmes very typically for 30 years. So, I could be quite certain that at least a third of the new century would, in general, be covered! Where did my future projection fall short? I had excluded from my abstract the role of that other probe of the structure of matter, electrons. In the decade 2010 onwards, major breakthroughs, or rather a collection of individual breakthroughs, occurred in recording and processing electron microscope images. One of these was dealing with the electron beam-induced blurring of the sample but which could be corrected by recording images in real time with an electronic detector. Sharp, much better resolved images could therefore be obtained. These allowed large biological molecular complexes to be resolved at the atomic level, and they didn't need to be crystallised. They do have to be very rapidly frozen first though and so the technique is called cryo-electron microscopy, abbreviated usually as 'cryoEM'. A second feature missing in my abstract was resolving structures at the atomic level at physiologically relevant temperatures, room temperature basically. Crystallography also had become largely done with cryo-cooled crystals. With non-damaging neutron beams used in crystallography and the expansion in capabilities, I had predicted with new instruments, software and methods a growing body of biological structures were available at both room temperature with the neutrons and their corresponding X-ray crystal structures at both cryo and room temperature. Comparing the structures of the same protein at room and cryo temperatures consistently showed types of differences. This knowledge also came from the time-resolved and dynamical crystallography studies done at the synchrotrons and then subsequently at the X-ray lasers. A new journal called *Structural Dynamics* became viable, launched by the American Institute of Physics Publishing and the American Crystallographic Association. I have been quite heavily involved with that, being on the Editorial Advisory Board and also chairing their Editor-in-Chief selection panel in 2020. So is that the end of cryo structural methods? No, not at all. The key thing is that a functional assay should accompany the cryo structural results. I mention a very effective one in Chapter 19 where my lab studied the blue colouration proteins extracted from the lobster shell. The cryo-cooled crystals were blue as they were at room temperature. Also the cryo crystallography protein structure gave a perfect calculated fit to the room temperature solution X-ray scattering measurements. Structures should hopefully also give predictions such as which are the functionally important amino acids. So, if these predicted-to-be-important amino acids are altered to a different amino acid by protein engineering, the protein function can be speeded up, slowed down or stopped. Such predictions where successful are testimony to the structure being true or not in all details.

My emphasis described above is obviously on experiments. As is evident elsewhere in this book, the critical importance of data and data archives are also vital to the scientific enterprise and the guarantee of its reproducibility and replicability.

I will come now to the other two abstracts, which are very focussed on the strength of computing.

Barry Robson and Jean Garnier, respectively, of the Unite de Bioinformatique, Biotechnologies, INRA, Jouy-en-Josas, France, and the J. T. Watson Research Center, IBM, New York, USA, are very careful in their predictions, referring only up to 2001:

Bioinformatics in the 21st century

Through 2001 (we) will see a breakdown of the artificial boundaries between the disciplines. ...Therefore one foresees researchers performing multiteraflop computation and handling petabytes of data through fast transmitting devices from multiple resources, with 3D-MG (three dimensional molecular graphics) involving high degrees of protein modelling.

I note, for example, their focus on the need to handle larger volumes of data; but they are reluctant to make any predictions as to different computational paradigms or, indeed, different scientific analytical techniques that could affect the type and volume of information that they might need to process in the future. One can say that they were wise to do so, given the very varied growth in this field in all directions over the last two decades.

The abstract from Charles Cantor CSO (Chief Scientific Officer), Sequenom, Inc., San Diego, and Sequenom, GmbH, Hamburg, Germany, is very expansive, and again selecting a portion from his abstract:

Biophysics will prosper in the 21st century

It was Walter Gibson, who first pointed out that biology was about to get a paradigm shift. He argued persuasively that the massive increase in available DNA sequence data would result in a decreased reliance on biological experimentation and an increased use of database searches and computer simulations that would either replace experiments or act as a prelude to them. As the 21st century dawns, we will see complete genome sequences for humans and many other species. We are also likely to see a complete enough survey of protein and nucleic acid motifs that computation will replace experiments in most macromolecular structure determination. The continuing increase in computer power available almost free to all biologists will drive a revolutionary change in the way scientific research and all other aspects of science are carried out... However the need for experiments will not end... the computer simulations will need scaling to measured in vitro reaction rate constants.

Take for example, the sentence of Charles Cantor 'The continuing increase in computer power available almost for free to all biologists will drive a revolutionary change in the way scientific research and all other aspects of science are carried out.' How can he have been so certain? The invention of the computer Colossus 2 took place at Bletchley Park in the later stages of the Second World War and was used by Alan Turing and co-workers to decode the encrypted messages of the German armed forces. Steadily, through the remainder of the 20th century, colossal changes in computer hardware took place. So, Charles Cantor could fairly reliably say the computer power and memory storage year on year improvements might indeed continue!

In conclusion, we scientists are actually rather careful in our predictions. That said other scientists would not, I know, have taken on presenting a talk in such an adventurously named session *Biophysics in the 21st Century*.

CHAPTER 24

Where Do Funding Agencies and Government Want to Take Us?

BY SPECIFYING FUNDING FOR particular objectives funding agencies and occasionally governments directly seek to shape the future of research. These are usually on fairly obvious, high-priority, societal relevant challenges. A current example is the coronavirus COVID-19 pandemic and how to solve it as expeditiously as possible includes a large component of science funding being directed at it (see Chapter 16).

More speculative are the 'moon shot' initiatives. The name stems from the announcement by the USA President John F Kennedy that the USA aimed to put a man on the moon before the end of the 1960s decade. The USA National Aeronautics and Space Administration (NASA) was charged with the task and accordingly provided a budget to achieve this aim. This was achieved on 20 July 1969 with Apollo 11 and accompanied by the famous words 'A giant leap for mankind'.

The UK Council for Science and Technology (CST) has provided detailed advice entitled 'Principles for science and technology moon-shots to achieve by 2030'.[1]

One has to say quite frankly that with so few politicians with any science training or knowledge top-down requests from Prime Ministers risk a colossal waste of resource whilst the researchers on the ground are scrabbling for small sums of money to take forward their ideas. However, the UK Government has, in parallel with the above moon shots initiative, launched a call for a national discussion of a Research and Development Roadmap by all stakeholders in the UK[2]:

Such 'brainstorming' discussions might deliver but I recall Max Perutz's answer[3] when asked by a Soviet Union Delegation to explain the success of the MRC Laboratory of Molecular Biology, of which he was Director at the time:

> They wanted to know how I had planned our successful Research Unit, imagining that I had recruited an interdisciplinary team as Noah had chosen the animals for his ark: two mathematicians, two physicists, two chemists, two biochemists and two biologists, and told them to solve the atomic structure of living matter. They were disappointed that the Unit had grown haphazardly and that I left people to do what happened to interest them.

But, that was then and this is now. Society is getting better at understanding science and technology. Maybe this is because as scientists we do get out and explain more and more what we do

---

[1] https://www.gov.uk/government/publications/principles-for-science-and-technology-moon-shots.

[2] https://www.gov.uk/government/publications/uk-research-and-development-roadmap. p. 58.

[3] https://www.nobelprize.org/prizes/uncategorized/the-medical-research-council-laboratory-of-molecular-biology.

in science to the public and to schoolchildren. Likewise Members of Parliament and Ministers are getting better at understanding science and technology. We should join in this growth in the desire by Government to discuss planning of such as a Research and Development Roadmap. In the UK, major venues for such discussions are the House of Commons and the House of Lords Science and Technology Committees [1,2]. One can make submissions of evidence directly as an individual or via one's professional organisation. An example of my own is at Ref. [3].

As a direct example of the public and schoolchildren showing an increased interest in science, including curiosity-driven science, I have worked with my University of Manchester Media Relations Officers on several occasions. An example can be found here [4]. It is through such efforts to explain our work to the public and school children that we can help our funding agencies have an easier job securing the future funding allocations, which are set overall by parliament and government.

REFERENCES

1. https://www.parliament.uk/business/committees/committees-archive/science-technology/.
2. https://www.parliament.uk/business/committees/committees-a-z/lords-select/science-and-technology-committee/publications/.
3. Helliwell, J.R. (2013). Evidence submission to the House of Lords Science and Technology Committee on *The implementation of open access* pp. 155–159 in https://www.parliament.uk/documents/lords-committees/science-technology/Openaccess/OpenAccessevidence.pdf.
4. Helliwell, J.R. (2004). Scientists' crystal clear view on drug revolution: A GROUND-breaking discovery by Manchester scientists is set to revolutionise the design of new drugs. https://www.manchestereveningnews.co.uk/news/greater-manchester-news/scientists-crystal-clear-view-on-drug-1129876.

Envoi

WHEN AND WHERE ARE ABOUT events, minor or major. In a posting I made on Facebook of the pages of Chapter 21 of my last book, *The Whats of a Scientific Life*, I had highlighted the merger of the International Council for Science (formerly the International Council of Scientific Unions, ICSU) with the International Social Sciences Council, the ISSC, to form the International Science Council. I stressed the good sense of this in the context of all the sciences assisting the United Nations in realising its Sustainability Development Goals (SDGs). A colleague posted in reply that the UN's SDGs were a 'dream'. I replied as follows:

> But what other tool besides its SDGs is there for the United Nations to make improvements by breaking those cycles of negative destiny, bad causes and miserable effects of poverty and unequal opportunity afflicting so many parts of the world?

I conclude my *Scientific Life* books with this assertion:

By means of our dreams, we can break causality, in science and in life, and make new and better things happen and find out new discoveries. We can strive for the paradise described in the poem by Coleridge that I quoted in the opening pages of this book.

Finally, dear readers, how to say Goodbye after these four books on the Scientific Life? As Shakespeare wrote in *Romeo and Juliet*:

Good night, good night! Parting is such sweet sorrow,
That I shall say good night till it be morrow.

PS My nominee for the first "When and Where of a Scientific Life" Award is:

The *Voyager* spacecraft 1 and 2, launched from Cape Canaveral Florida by the Jet Propulsion Laboratory (JPL) in Los Angeles California in 1977, and which left the solar system in 2012, sending back data to JPL well beyond their initial 5 year lifetimes [1]. These data are declared by JPL "to continue until around the year 2020 when the spacecraft's ability to generate adequate electrical power for continued science instrument operation will come to an end" [1].

REFERENCE

1. https://voyager.jpl.nasa.gov/mission/science/planetary-voyage/.

Bibliography

Helliwell, J.R. (2017). *Skills for a Scientific Life*, CRC Press, Boca Raton, FL.

Helliwell, J.R. (2019). *The Whys of a Scientific Life*, CRC Press/Taylor & Francis, Boca Raton, FL.

Helliwell, J.R. (2020). *The Whats of a Scientific Life*, CRC Press/Taylor & Francis, Boca Raton, FL.

Samuel Taylor Coleridge. (1772–1834). https://poets.org/poem/kubla-khan.

Name Index

Subject Index